Internet of Things

Technology, Communications and Computing

Series editors

Giancarlo Fortino, Rende (CS), Italy
Antonio Liotta, Eindhoven, The Netherlands

More information about this series at http://www.springer.com/series/11636

Beniamino Di Martino · Kuan-Ching Li
Laurence T. Yang · Antonio Esposito
Editors

Internet of Everything

Algorithms, Methodologies, Technologies and Perspectives

Springer

Editors
Beniamino Di Martino
Department of Industrial and Information Engineering
Università degli studi della Campania Luigi Vanvitelli
Naples
Italy

Kuan-Ching Li
Providence University
Taichung
Taiwan

Laurence T. Yang
Department of Computer Science
St. Francis Xavier University
Antigonish, NS
Canada

Antonio Esposito
Università degli studi della Campania Luigi Vanvitelli
Aversa
Italy

ISSN 2199-1073 ISSN 2199-1081 (electronic)
Internet of Things
ISBN 978-981-10-5860-8 ISBN 978-981-10-5861-5 (eBook)
https://doi.org/10.1007/978-981-10-5861-5

Library of Congress Control Number: 2017947449

Printed on acid-free paper

This Springer imprint is published by Springer Nature
The registered company is Springer Nature Singapore Pte Ltd.
The registered company address is: 152 Beach Road, #21-01/04 Gateway East, Singapore 189721, Singapore

Contents

Introduction

Connected sensors and devices are a pervasive reality, as we interact more or less knowingly with smart items everyday. While smartphones represent the main Internet-connected devices which people interact with during their daily lives, other intelligent items are at our disposal, and they are so integrated with the environment that we fail to notice them. Animals too participate in this relatively new phenomenon, as the use of subcutaneous chips to identify and track them are widely used. In the modern era, where each item and living being (animal or human) are or can be connected to others and share data, the Internet of Everything has become a new buzzworld, and great attention has aroused for the huge variety of possible applications of such technologies, together with concerns regarding privacy and security.

Previous Springer books have been published on the Internet of Things topics, such as [1] and [2], the latter belonging to the same **Internet of Things** series of the present volume. In the present book, several aspects of the IoT phenomenon are analyzed, corresponding challenges are pointed out, and solutions or practical suggestions are proposed. The general organization of the book is as follows:

1. **An Introduction to the Internet of Everything** provides an introduction to the concept of Internet of Everything and focuses on some of the specific technological areas which fall under this wide category.
2. **Chapter 2**: Towards an Integrated Internet of Things: Current Approaches and Challenges introduces the main technologies currently available to define a machine readable and human comprehensible IoT API and points out the several challenges which will derive from an automatic analysis and description of IoT interfaces.
3. **Chapter 3**: Energy Harvesting in Internet of Things provides a comprehensive review of IoT devices, from their roles and responsibilities, to the challenges of operating them autonomously in heterogeneous environment
4. **Chapter 4**: A Detailed Analysis of IoT Platform Architectures: Concepts, Similarities, and Differences conducts a detailed analysis of several

state-of-the-art IoT platforms in order to foster the understanding of the underlying concepts, similarities, and differences between them.
5. **Chapter 5**: Fog Computing: A Taxonomy, Survey and Future Directions presents a taxonomy of Fog computing according to a set of identified challenges and its key features. Also, it maps the existing works to the taxonomy in order to identify current research gaps in the area of Fog computing.
6. **Chapter 6**: Challenges and Opportunities in Designing Smart Spaces takes a comprehensive look at the challenges in developing user-centric smart spaces for two different smart space scenarios: Smart Home and Smart Shopping.
7. **Chapter 7**: SMART-FI: Exploiting Open IoT Data from Smart Cities in the Future Internet Society introduces a novel Smart City platform being developed in the context of SMART-FI project, which aims to facilitate analyzing, deploying, managing, and interoperating Smart City data analytics services.
8. **Chapter 8**: A Case for IoT Security Assurance discusses and analyzes challenges in the design and development of assurance methods in IoT, focusing on traditional CIA properties, and provides a first process for the development of continuous assurance methods for IoT services.
9. **Chapter 9**: Study on IP Protection Techniques for Integrated Circuit in IOT Environment focuses on Intellectual Property (IP) protection and in particular how to hide secrets into IP circuit and authenticate IP via the secrets.
10. **Chapter 10**: Cyber Defense Capabilities in Complex Networks focuses on cyber security issues in networks and provides a study on a real scenario.

References

1. Fortino, G., and P. Trunfio. 2014. *Internet of Things Based on Smart Objects*. Springer, Berlin. doi:10.1007/978-3-319-00491-4
2. Guerrieri, V. Loscri, A. Rovella, and G. Fortino. 2016. *Management of Cyber Physical Objects in the Future Internet of Things. Internet of Things*. Springer, Berlin. doi:10.1007/978-3-319-26869-9

Trends and Strategic Researches in Internet of Everything

Beniamino Di Martino, Kuan-Ching Li, Laurence Tianruo Yang and Antonio Esposito

Abstract Connected sensors and devices are a pervasive reality, as we interact more or less knowingly with smart items every day. While smart-phones represent the main Internet connected devices which people interact with during their daily lives, other intelligent items are at our disposal and they are so integrated with the environment that we fail to notice them. Animals too participate in this relatively new phenomenon, as the use of subcutaneous chips to identify and track them are widely used. In the modern era, where each item and living being (animal or human) is or can be connected to others and share data, the Internet of Everything has become a new buzz-world, and great attention has aroused for the huge variety of possible applications of such technologies, together with concerns regarding privacy and security. In this chapter we introduce the concept of Internet of Everything and we focus on some of the specific technological areas which fall under this wide category.

1 The Internet of Everything

Due to recent advancements in connection technologies and the widespread of smart devices, capable to communicate and exchange consistent loads of information, our environment is transforming into an "**Internet of Everything**"(IoE). The Internet

---

B. Di Martino (✉) · A. Esposito
Department of Industrial and Information Engineering, Università degli Studi della Campania Luigi Vanvitelli, Via Roma 29 Aversa (CE), Caserta, Italy
e-mail: antonio.esposito@unicampania.it

B. Di Martino
e-mail: beniamino.dimartino@unina.it

K.-C. Li
Department of Computer Science and Information Engineering, Xiamen University, Software School, China and Providence University, Taichung, Taiwan
e-mail: kuancli@gm.pu.edu.tw

L.T. Yang
Department of Computer Science, St. Francis Xavier University, Antigonish, NS B2G 2W5, Canada
e-mail: ltyang@gmail.com

B. Di Martino et al. (eds.), *Internet of Everything*, Internet of Things,
https://doi.org/10.1007/978-981-10-5861-5_1

of Everything has become a catch-all phrase to describe adding connectivity and intelligence to just about every device in order to give them special functions. However, this can be quite reductive, as IoE provides links not only among things, but also data, people and (business) processes. Evolution of current sensor and device networks, with strong interaction with people and social environments, will have a dramatic impact on everything from city planning, first responders, military, and health. Several Internet and connection-based paradigms fall under the IoE umbrella, such as:

- **Internet of Thing** (IoT), which connects cyber, physical and biological worlds via smart sensors and devices strongly immersed in the surrounding environment.
- **Internet of People** (IoP), which considers links and connections among people, and their social interactions.
- **Industrial Internet** (II), more focused on data of interest for/coming from Industries.

Not surprisingly at all, early attempts and deployments of Internet of Things networks began with connecting industrial equipment. However, today, the vision of IoT has expanded to connect everything from industrial equipment to everyday objects. Not only mechanical and electronic devices can be connected together in an IoT environment: this can include living organisms such as plants, farm animals and, obviously, people. For example, an interesting project in Essex (UK), the **Cow Tracking Project**, has connected cows to the Internet via radio positioning tags, to monitor them for illness, track their behaviour in the herd and detect abnormal movements and actions. Academic research on animal wearables has also pointed out in this direction [14].

Wearable computing and digital health devices, such as **Nike+ Fuel** [20] band and **Fitbit** [9], are examples of how people are connecting in the Internet of Things landscape, and have been subjected to all kinds of possible tests and applications [24, 26]. The extension of the IoT definition to include people, places, objects and things, and the coining of the Internet of Everything term can be tacked back to Cisco [7]. Basically anything you can attach a sensor to and provide connectivity can participate in the new connected ecosystems.

While such areas cover many aspects of today's life, there is still the strong need to contextualize and integrate data and information coming from different networks and frameworks. Indeed, there is the need to provide a common ground for integrating information coming from heterogeneous sources. Such a shared ecosystem would allow for the interaction among data, sensor inputs and heterogeneous systems. In order to enable such an integrated framework, semantics become a fundamental component: semantic technologies are able to provide the necessary bridge between different data representations, and to solve terminology incongruence.

2 The Internet of Things

The term Internet of Things (IoT) has been introduced by industry researchers, but it has rapidly emerged into mainstream public view, thus becoming a notorious "buzz word ". In a few words, IoT represents the ability of network devices to sense, collect and sometimes even analyse in loco data from the world around us, and then share that data across the Internet. In [10] the authors present a series of architectures, algorithms, and applications for Internet of Things based on Smart Objects, providing an interesting overview of the IoT domain.

Sharing is one of the main feature of IoT devices, as only in this way data can be efficiently processed and utilized for various interesting purposes. There are discordant opinions regarding the impact of IoT on our future lives: some claim it will completely transform how computer networks are used, thus having a deep impact on future developments of the IT industry; others simply believe that the IoT hype will not have much impact into the daily lives of most people, and that it will fade away with time.

Whichever will be the future of IoT, as of now its applications are numerous and outstanding:

- Complex proximity systems can be used to build self-parking cars;
- **Wearable devices** and phones can be used to track people's movement, exercise habits and day-to-day activities. All these information can be then exploited for athletes' goal tracking and physical activities planning, or to simply track people and identify their position whenever needed. Figure 1 show some of the most common wearable devices typologies;
- Vehicle tracking, via ad-hoc GPS devices or common smartphones, is a widespread method for determining the position, as an instance, of delivery trucks, or to determine traffic congestions in specific areas.
- Wearable devices, equipped with special sensors, can detect physical dangers and communicate it to the wearers or to people in the proximity, via the Internet.
- Domotics appliances can exploit combined sensors' data to provide real **Smart Home** experiences. Devices that automatically tune indoor temperature according to people demands and environmental conditions can be already found in many homes, while systems that can automatically order online groceries and home supplies are being developed and tested.

As stated before, the ability to connect to the Internet and among them is a fundamental requirement of IoT devices. As of today, many of the most common house devices with which we continuously interface and interact, can be modified to work in an IoT system. Dishwashers, washing machines, heat pumps are only a few of the households that can be *upgraded* with a selection of Wi-Fi network adapters, motion sensors, cameras, microphones and other instrumentation which would enable them to work in an IoT system.

Smaller and more primitive versions of such kind of devices already exist in common houses: just thing of intelligent light bulbs, which are turned on and off

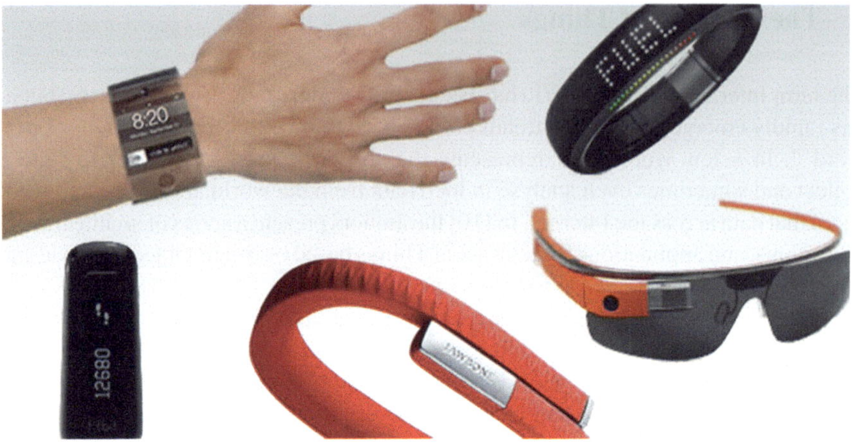

Fig. 1 Common wearable devices

according to the movements of people detected by sensors inside and outside rooms. Such simple devices have already been adopted in many houses and public structures, to reduce energy consumption. Wireless scales and blood pressure monitors represent early examples of IoT gadgets. Smart watches and glasses will play a key role in future IoT systems, as they represent a natural extension of very common wearable items. Also, the diffusion of Wi-fi connections and the availability of Bluetooth technology on almost every device, guarantees the required connection among sensors and with the Internet.

The intrinsic characteristics of IoT make it very different from traditional computing. Data with which IoT devices deal are generally very small n size, but they are frequently, if not continuously, transmitted. Often, it is necessary to talk about **Data Streams**, as such information flow without interruption and need to be taken care of rapidly. The number of devices and computing nodes, concurrently connected to the network and communicating through it are much greater in IoT than in any other computing paradigm. The work presented in [13] covers a wide range of topics related to smart devices, such as resource management, hardware platforms, communication and control, and control and estimation over networks. It also discusses decentralized, distributed, and cooperative optimization as well as effective discovery, management, and querying of Cyber Physical Objects (CPOs).

2.1 Issues Around IoT

Considering the wide range of subjects touched by the IoT development, it is obvious that many concerns and issues have arisen regarding different topics.

The first and more immediate questions arisen by IoT regard the **Privacy of personal data**. Wearables, smartphones and other Internet connected devices deliver a lot of information about ourselves: physical location, updates about our weight and blood pressure (that have been made accessible by our health care providers), our relationships with people we are interconnected to, and more detailed data about ourselves are continuously streamed over wireless networks and potentially around the world. Means to protect such data from malicious sniffing are being researched and developed, but educating people to understand what they actually share and the potential issues related to information disclosure is a strong topic.

Connected devices require energy to work, even more if they are wireless due to the power necessary to sustain communications. However, supplying power to this new proliferation of IoT devices and their network connections can be expensive and logistically difficult. Portable devices require batteries that need to be periodically recharged, or to be eventually replaced. Even devices optimized for lower power usage need energy to work, and the cost to supply such power to billions of different electrical components is potentially huge.

New Business opportunities have been disclosed by the advent of IoT technologies and, as a consequence, numerous corporations, start-ups and joint ventures have born. Competition in the market surely concurs to lower prices for customers, but it also generates a huge number of offers, products and services which, in many cases, lack interoperability capabilities and tend to confuse the consumers.

Another issue regards the actual availability of an efficient network to allow communication among distributed devices. IoT assumes that network equipment and network-based technologies can interoperate intelligently and automatically, but sometimes even guaranteeing a stable connection to mobile sensors can be a challenging task. Also, the IoT network should be able to adapt to the ever changing needs of consumers, who could interact with the connected devices in different, and often unpredictable, ways. The Internet-of-Things vision provides a large set of opportunities to users, manufacturers and companies [19]. In fact, IoT technologies will find wide applicability in many productive sectors including, e.g., environmental monitoring, health-care, inventory and product management, workplace and home support, security and surveillance. From a user point of view, the IoT will enable a large amount of new always responsive services, which shall answer to users needs and support them in everyday activities. Before the IoT and big data coalition can deliver on its promise, there are a number of barriers to overcome.

The first challenge is the worldwide adoption of shared standards. The use of standards ensures interoperable and cost-effective solutions, opens up opportunities in new areas and allows the market to reach its full potential. For the IoT to work, there must be a framework within which devices and applications can exchange data securely over wired or wireless networks. In this area there are a lot of player: M2M [21], AllJoyn [2], OIC [22].

M2M (Machine to Machine) refers to technologies that allow both wireless and wired systems to communicate with other devices of the same type. M2M is a broad term as it does not pinpoint specific wireless or wired networking, information and communications technology. This broad term is particularly used by business executives. M2M is considered an integral part of the Internet of Things (IoT) and brings several benefits to industry and business in general as it has a wide range of applications such as industrial automation, logistics, Smart Grid, Smart Cities, health, defense etc. mostly for monitoring but also for control purposes. Open Machine to Machine (OM2M) [1] provides an open source service platform for M2M interoperability based on the ETSI-M2M standard. OM2M follows a RESTful approach with open interfaces to enable developing services and applications independently of the underlying network. It proposes a modular architecture making it highly extensible via plugins. It supports multiple protocol bindings such as HTTP and CoAP. Various interworking proxies are provided to enable seamless communication with vendor-specific technologies such as Zigbee and Phidgets devices. OM2M implements the SmartM2M [17] standard. It provides a horizontal Service Capability Layer (SCL) that can be deployed in an M2M network, a gateway, or a device. Each SCL provides Application Enablement, Generic Communication, Reachability, Addressing and Repository, Interworking proxy, Entity Management, etc. It includes several primitive procedures to enable machines authentication, resources discovery, applications registration, containers management, synchronous and asynchronous communications, access rights authorization, groups organisation, re-targeting, etc.

The AllSeen Alliance [2] is a cross-industry consortium dedicated to enabling the interoperability of billions of devices, services and apps that comprise the Internet of Things. AllJoyn,a framework created by the AllSeen Alliance, is an open, universal, secure and programmable software connectivity and services framework that enables companies and enterprises to create interoperable products that can discover, connect and interact directly with other AllJoyn-enabled products. AllJoyn is agnostic respect to the transport layer, the OS, the platform and brand, enabling the emergence of a broad ecosystem of hardware manufacturers, application developers and enterprises that can create products and services that easily communicate and interact. It consists of an open source SDK and code base of service frameworks that enable such fundamental requirements as discovery, connection management, message routing and security, ensuring interoperability among even the most basic devices and systems. The initial planned set of service frameworks include: device discovery to exchange information and configurations (learning about other nearby devices); onboarding to join the users network of connected devices; user notifications; a common control panel for creating rich user experiences; and audio streaming for simultaneous playback on multiple speakers. In addition, the Alliance is producing developer tools and verifying correct implementation through a compliance program.

The Open Interconnection Consortium (OIC) is defining a common communication framework based on industry standards to wirelessly connect and manage the flow of information among IoT devices. It sponsors the IoTivity Project [15], an open source software framework for device-to-device connectivity. The IoTivity architectural goal is to create a new standard by which billions of wired and wireless devices will connect to each other and to the internet. The goal is an extensible and robust architecture that works for smart and thin devices. The IoTivity framework APIs expose the framework to developers, and are available in several languages and for multiple operating systems. The APIs are based on a resource-based, RESTful architecture model. The framework operates as middleware across all operating systems and connectivity platforms and has four essential building blocks: discovery, data transmission, data management and device management. IoTivity Services, which are built on the IoTivity base code, provide a common set of functionalities to application development. IoTivity Services are designed to provide easy and scalable access to applications and resources and are fully managed by themselves. There are six IoTivity Services in v1.0, each with its own unique functionality: Resource Encapsulation, Resource Container, Things Manager, Resource Hosting, Resource Directory and MultiPHY EasySetup. Resource Encapsulation abstracts common resource function modules. It provides functionalities for both the client and server side functions to IoTivity Service developers. For client side, it provides resource Cache and Presence Monitoring functions. On the other hands, for the server side, it provides the simple, direct way to create the resource and to set the properties, attributes of resources. Resource Container provides a way to integrate non-OIC resources into OIC ecosystem by creating, registering, loading and unloading resource bundles. It also provides common resource templates and configuration mechanism for resource bundles. It deals with OIC specific communication features, and provides common functionalities in a generic way. Things Manager creates Groups, finds appropriate member things in the network, manages member presence, and makes group action easy. The goal of Resource Hosting is to off-load the request handling works from the resource server where original resource is located to reduce the power consumption of resource constrained devices. A resource directory is a server that acts on behalf of the thin-client. The thin-client after it publishes their resources, resource-directory will respond on behalf of these devices. The device acting as a resource directory could itself hold resources. MultiPHY Easysetup is an IoTivity primitive service to enable different sensor devices(with different connectivity support) to be easily connected to the end user's IoTivity network seamlessly. Thus enabling Sensor devices to be part of the IoTivity network in a user friendly manner. The big data universe can provide infrastructure and tools for handling,processing and analysing deluge of the IoT data. However, there is a lack of efficient methods and solutions that can structure, annotate, share and make sense of the IoT data and facilitate transforming it to actionable knowledge and intelligence in different application domains [3]. The issues related to interoperability, automation, and data analytics naturally lead to a semantic-oriented perspective towards IoT. Semantic technologies based on machine-interpretable representation formalism have shown promise for describing objects, sharing and integrating information, and inferring new knowledge together with other intelligent processing techniques.Semantic technologies are necessary to integrate data from heterogeneous data sources. In fact ontologies and related semantic technologies, such as ontology merging and mapping, could offer a simple and powerful mean to provide not only a formats unifier but also a semantic translator by providing an unified interpretation of different data sources. Moreover, by means of semantic annotation the data will be self explanatory information carriers and thus enabling the dynamic discovery of relevant data sources and data. Semantic technologies enable the development of extensible context models which can be adapted to different application domain and that can be easily enriched to accommodate the continuous evolution of IoT systems. There are many efforts in creating common models for describing and representing the IoT data and resource descriptions, among many IOT-A [23], SSN [6], OpenIoT [18].

3 Industrial Internet

The term **Industrial Internet** (II) has been used for the first time in 2012 by **General Electric**. It is currently used as a synonymous of Industrial Internet of Things or IIoT. Since then, many other companies have shown great interest in II, as they have invested in researches in this field: among them, noticeable examples are Googe, Cisco or DataLogic.

The Industrial Internet Consortium [25], which was founded by prominent industrial actors such as AT&T, Cisco Systems Inc., General Electric, IBM and Intel, also advocates for the advancement of the Industrial IoT. The Consortium is a not-for-profit organization that manages and advances the growth of the Industrial IoT through the collaborative efforts of its member companies, industries, academic institutions and governments. In the years, it has organized events to sponsor the latest technologies and research in the II field, by also providing test-beds for new solutions. As for th more generic Internet of Things scenario, the Industrial Internet implies a collection of numerous devices, connected by communications software and cabled or wireless networks. The resulting frameworks and the individual devices that compose it, are able to collect and almost instantly act on information, by also modifying their behaviour and actively interacting with the surrounding environment. The key difference between IoT and II is represented by human interactions: in a real II scenario, human intervention is not required at all. Indeed, II solutions are widely adopted whenever human cannot interact with or operate directly in the target environment. Space vectors for long range explorations or dangerous expeditions, or mining robots for deep excavations do not require to be operated by humans, but can take smart and fast decisions according to the immediate information they receive from the surroundings.

Another key point is therefore represented by the criticality of almost all of the Industrial IoT applications: aerospace and defence, healthcare and energy, are all fields in which a system failure can result in life-threatening or other emergency situations. Instead, IoT systems tend to be consumer-level devices, which still represent valid and important commodities, but eventual breakdowns do not immediately create emergency situations.

There are many practical examples of II applications which are worth mentioning

- Vast networks of connected sensors can be employed to monitor Oil pipelines, in order to detect damages and promptly act to fix them and avoid spilling.
- Intelligent railway collision avoidance systems exploit data coming from GPS antennas, installed on trains, to determine locomotives' positions and avoid accidents

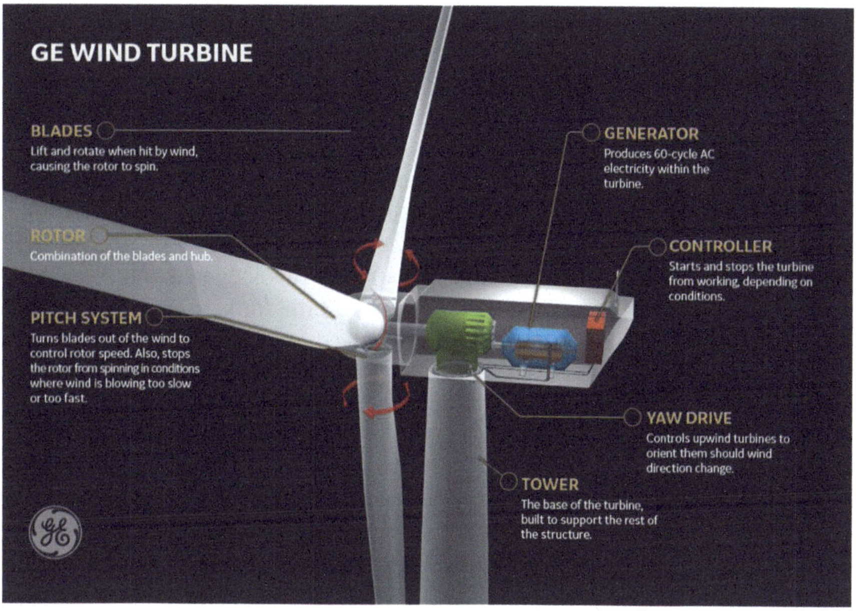

Fig. 2 Schema of a Smart Wind Turbine from General Electric

- Water quality sensors can be used to check tap water distributed to citizens, determine and avoid leaks and wastes.
- The renewable energy sector can also benefit from advances sensor networks: Digital Wind Farms [11] exploit data captured by sensors regarding the surroundings to determine which turbine to turn of and off in order to best exploit wind strength, or to signal failures in one or more components of a single turbine. Figure 2 reports the schema of a Wind Turbine designed b General Electric.

4 The Internet of People

We are all accustomed to hear from the news, magazines or scientific reports, of the billions of connected things we are continuously surrounded by. But what are those things? They surely are not devices only: human assets are a part of such an extended network and can be managed through technology.

While the Industrial Internet is based on the idea that human intervention is not needed, in an Internet of People scenario humans represent the focus. The ultimate objective of interconnected sensors and devices in this context is to **anticipate the owner's needs**: an umbrella could advice a person to bring it with him, if it will rain; a smart fridge could automatically order supplies or advice the owner to throw away expired food; smart bins could help citizen in recycling.

Traditional IoT devices, in particular wearables (Smart bands, smart glasses), often propose an interaction model that requires the use of windows and screens: this is a not natural way, for a human being, to approach the surrounding world. People interact with objects by using their own body, so interposing a device which is not fully blended with the environment somehow disrupts the whole idea behind the IoP concept. In theory, people should not be even aware of their interaction with a machine. Obviously, this can arise many ethical issues, as an abuse of such technologies could make people too dependant on them and could de-personalize many aspects of everyday life [16]. Indeed, current technology makes it possible to move the integration between man and machines a little further: implantable chips are a reality, and their use has been already approved in some public environments, such as hospitals [8] and more recently in private offices [4]. The potential uses of such a technologies are practically unlimited: secure access to products and services, automatic payments via a direct connection to a bank account, tracking, people recognition. Having all your medical records saved on such a chip, immediately ready for doctors in case of emergency, can become very useful.

As already stated before, strong privacy issues arise when data about people are shared via a network of connected devices or the Internet. Generally, the main focus is represented by the **location** of a person, which is regarded as one of the most sensitive information being involved in many IoT base applications. Just consider the recent hype in the industry of location based games, such as **Ingress** [5] and the very notorious **Pokemon-Go** [12]. In such games people use their smart devices (not only phones, as dedicated smart-watches and bands have been designed just to interact with them) to interact with an augmented-reality world. However, as interaction with the environment requires and exact localization of a person and the notification of her position to other players (not considering the data collected by the game servers about your movements), security issues are bound to rise. Especially if we consider that among the users of such applications minors represent a consistent percentage. However, a clear distinction between tracking people for a specific benevolent application cannot be mixed with snooping into people's private lives. Tracking the whereabouts of a person can indeed represent a good solution to many problems for which there is no other satisfactory approaches:

- Constantly monitor the position of workers that operate in hazardous environments, such as miners, rescuer teams, trekkers, Alpinists and similar;
- Being aware of the location of children, during routes to school, school hours or on their way home;
- **Mobile Personal Emergency Response Systems** (mPERS) represent an excellent use of GPS tracking. By knowing their exact location, people in distress can be immediately succoured.
- Penitentiary facilities can exploit location based service to locate prisoners in case of escape, or to control their movements within the facilities.

Obviously, location is only one of the sensitive information that Internet connected devices can disclosure, if not well used or if a malicious actor interferes with their regular functioning. Personal data, such as home addresses, phone numbers, or credentials for the access to bank account and credit cards, can be inadvertently shared on a devices network, and be consequently sniffed and used for illegal purposes.

5 Conclusion

In this chapter we have introduced the main concepts behind the phenomena of the Internet of Things (IoT), Industrial Internet (II) and Internet of People (IoP), which together build the pillars of the Internet of Everything (IoE) ecosystem. We have stressed the importance of network availability to enable an efficient device network and pointed out both technical issues (such as the availability of a proper power supply for the connected devices) and security concerns (disclosure of location and other sensitive data). This does not exhaust the subject: on the contrary, each of the points we have merely touched in this introductory chapter deserves to be deepened and studied in the proper context.

References

1. Alaya, M. Ben, Yassine Banouar, Thierry Monteil, Christophe Chassot, and Khalil Drira. 2014. Extensible etsi-compliant m2m service platform with self-configuration capability. *Procedia Computer Science* 32: 1079–1086.
2. AllJoyn. https://allseenalliance.org/.
3. Barnaghi, Payam, Wei Wang, Cory Henson, and Kerry Taylor. 2012. Semantics for the internet of things: Early progress and back to the future. *International Journal on Semantic Web and Information Systems (IJSWIS)* 8(1): 1–21.
4. Cellan-Jones, Rory. 2015. Office puts chips under staffs skin. *BBC News*.
5. Chess, Shira. 2014. Augmented regionalism: Ingress as geomediated gaming narrative. *Information, Communication and Society* 17(9): 1105–1117.
6. Compton, Michael, Payam Barnaghi, Luis Bermudez, RaúL GarcíA-Castro, Oscar Corcho, Simon Cox, John Graybeal, Manfred Hauswirth, Cory Henson, Arthur Herzog, et al. 2012. The ssn ontology of the w3c semantic sensor network incubator group. *Web Semantics: Science, Services and Agents on the World Wide Web* 17: 25–32.
7. Evans, Dave. 2012. The internet of everything: How more relevant and valuable connections will change the world. *Cisco IBSG*, 1–9.
8. Feder, Barnaby J, and Tom Zeller, Jr. 2004. Identity badge worn under skin approved for use in healthcare. New York Times.
9. Fitbit official site for activity trackers and more. https://www.fitbit.com/. Accessed Jan 2017.
10. Fortino, Giancarlo, and Paolo Trunfio. 2014. *Internet of things based on smart objects*. Springer.
11. GE digital wind farm. www.gerenewableenergy.com/wind-energy/technology/digital-wind-farm.html. Accessed 20 Mar 2016.
12. Gregory, Brent, Sue Gregory, and Boahdan Gregory. Harvesting the interface: Pokémon go. In *33rd international conference of innovation, practice and research in the use of educational technologies in tertiary education*, 240.

13. Guerrieri, Antonio, Valeria, Loscri, Anna, Rovella, and Giancarlo, Fortino. 2016. *Management of cyber physical objects in the future internet of things*. Internet of things. Springer International Publishing.
14. Huhtala, A., K. Suhonen, P. Mkel, M. Hakojrvi, and J. Ahokas. 2007. Evaluation of instrumentation for cow positioning and tracking indoors. *Biosystems Engineering* 96(3): 399–405.
15. IoTivity. https://www.iotivity.org/.
16. Judge, J, and J. Powles. 2015. Forget the internet of things we need an internet of people. https://www.theguardian.com/technology/2015/may/25/forget-internet-of-things-people. Accessed 20 Mar 2016.
17. Kanti Datta, Soumya, and Christian Bonnet. 2014. Smart m2m gateway based architecture for m2m device and endpoint management. In *Internet of Things (iThings), 2014 IEEE international conference on, and green computing and communications (GreenCom), IEEE and Cyber, Physical and Social Computing (CPSCom), IEEE*, 61–68. IEEE.
18. Kim, Jaeho, and Jang-Won Lee. 2014. Openiot: An open service framework for the internet of things. In *2014 IEEE World Forum on, Internet of things (WF-IoT)*, 89–93. IEEE.
19. Miorandi, Daniele, Sabrina Sicari, Francesco De Pellegrini, and Imrich Chlamtac. 2012. Internet of things: Vision, applications and research challenges. *Ad Hoc Networks* 10(7): 1497–1516.
20. Nike+ Fuel band. http://www.nikeplus.com.br/. Accessed Jan 2017.
21. Niyato, Dusit, Lu Xiao, and Ping Wang. 2011. Machine-to-machine communications for home energy management system in smart grid. *Communications Magazine, IEEE* 49(4): 53–59.
22. Open connectivity foundation. http://openconnectivity.org/.
23. Responsible Beneficiary, IML FhG, Stephan Haller SAP, Edward Ho HSG, Christine Jardak, Alexis Olivereau CEA, Alexandru Serbanati, Matthias Thoma SAP and Joachim W Walewski. Internet of things-architecture iot-a deliverable d1. 3–updated reference model for iot v1. 5.
24. Takacs, Judit, Courtney L Pollock, Jerrad R Guenther, Mohammadreza Bahar, Christopher Napier, and Michael A Hunt. Validation of the fitbit one activity monitor device during treadmill walking. *Journal of Science and Medicine in Sport* 17(5): 496–500.
25. The industrial internet consortium. http://www.iiconsortium.org/. Accessed 20 Mar 2016.
26. Tucker, Wesley J., Dharini M. Bhammar, Brandon J. Sawyer, Matthew P. Buman, and Glenn A. Gaesser. 2015. Validity and reliability of nike+ fuelband for estimating physical activity energy expenditure. *BMC Sports Science, Medicine And Rehabilitation*, 7(1): 14.

Towards an Integrated Internet of Things: Current Approaches and Challenges

Beniamino Di Martino, Antonio Esposito, Stefania Nacchia and Salvatore Augusto Maisto

Abstract With the diffusion of sensors and smart devices, and the advances in connection technologies, the Internet of Things (IoT) has become a very popular topic. Because of the creation and expansion of new and existing sensor networks, the need to define a common standard for sensors' interfaces representation has arisen. Currently it is difficult to make different sensors and sensors' networks interoperate seamlessly, since their interfaces are not always well specified or are not ready to be adapted immediately to one another. In this chapter we will introduce the main technologies currently available to define a machine readable and human comprehensible IoT API, and we will point out the several challenges which will derive from an automatic analysis and description of IoT interfaces. Security issues are also considered and discussed.

1 Introduction

The amount of devices that connect to the Internet and exchange data among them or share it to the world grows day by day. The IoT phenomenon generates unprecedented amounts of data, with peculiar characteristics, as they are generally represented by small data chunks which travel fast and are continually streamed through networks to and from devices and sensors. This has strong impact on the amount of computational power needed, not only to elaborate but also to simply manage them. As

B. Di Martino · A. Esposito (✉) · S. Nacchia · S.A. Maisto
Department of Industrial and Information Engineering,
Università degli Studi della Campania Luigi Vanvitelli, Via Roma 29 Aversa (CE),
Aversa, Italy
e-mail: antonio.esposito@unicampania.it

B. Di Martino
e-mail: beniamino.dimartino@unina.it

S. Nacchia
e-mail: stefania.nacchia@unicampania.it

S.A. Maisto
e-mail: salvatoreaugusto.maisto@unicampania.it

B. Di Martino et al. (eds.), *Internet of Everything*, Internet of Things,
https://doi.org/10.1007/978-981-10-5861-5_2

a consequence, The IoT and Big Data universes are considered as two inseparable aspects of modern IT evolution, and are set to transform many areas of business and everyday life. IoT has brought new opportunities, both for users and manufacturers [22], and new businesses were born as a consequence. Many productive sectors will benefit from the application of IoT technologies, ranging from environmental monitoring, to health-care and security and surveillance.

However, despite the great interest around IoT and related technologies, there are still several issues that need to be solved, and that limit the adoption of current solutions worldwide. Surely, the first and more challenging issue to be faced regards the adoption of a common and shared standard for the description of sensors' and devices' interfaces. As it often happens with rapidly developing IT technologies, the market has already been flooded with new offers, services and technological solutions: however, since the proposed products hardly share common programming interface in form of standardized APIs, Interoperability becomes extremely difficult. Also, even expert programmers can encounter difficulties when dealing with such a huge variety of ever evolving services and interfaces.

In this manuscript a general model for an IoT framework will be provided, with the intention to identify the main strategies which can lead to the solution of interoperability issues among IoT products and services, by also analysing pros and cons of the provided solutions.

The reminder of this paper is organized as follows: Sect. 2 provides an overview of current research projects which aim at defining new shareable formalisms for IoT APIs description and of current technologies in the field; Sect. 3 introduces the problem of representing and sharing IoT APIs, and describes and compares the main formalisms currently available on the market; Sect. 5 points out the main challenges in automatic API analysis; Sect. 6 stresses the importance of security when interacting with RESTful interfaces and points out the unreliability of current approaches; finally, Sect. 7 closes the chapter with comments and considerations on the presented work.

2 State of the Art

In the latest years new technologies as IoT, smart grid, smart appliances, and wearable device powered health and fitness have been emerging as major application domains but with varying architecture and data models. The high availability of new sensors, with different and precision capabilities and fields of application, has made possible to collect huge volumes of data that can be elaborated and combined to provide valuable information. The **Health Care Domain** in particular has been influenced by such innovations. The work presented in [14] provides a survey discussing clear motivations and advantages of multi-sensor data fusion, and particularly focuses on physical activity recognition, aiming at providing a systematic categorization and common comparison framework of the literature. Figure 1 shows vertical silos for these domains with examples including physical sensors to the Internet service.

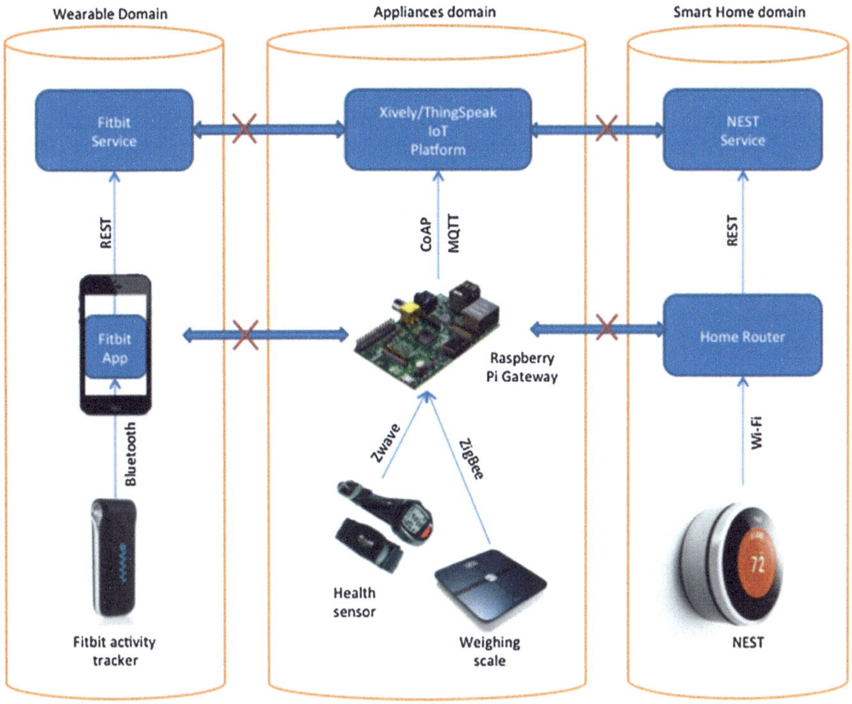

Fig. 1 IoT services architecures

In the health care domain **Fitbit**, an activity monitoring device, provides complete sets of IoT components creating proprietary and closed silo. It also provides graphical interfaces and uses a RESTful application interface to connect the sensors to their cloud service. Similarly, a user can connect and monitor his health by analysing data from sensors such as heart rate, glucose, weighing scale using any popular open hardware platform such as Raspberry Pi or Arduino as a gateway node and with an ad hoc web service that grants access to the data through a RESTful interface.

In [9] the authors describe **Body Sensor Networks** (BSNs) as a collection of wearable (programmable) sensor nodes communicating with a local personal device. The sensor nodes have computation, storage, and wireless transmission capabilities, a limited energy source, and different sensing capabilities. In order to avoid one of the most limiting issues in the adoption of BSNs, that is the complexity of their programming interfaces, the authors have developed the **Signal Processing In Node Environment (SPINE)**, an open-source programming framework, designed to support rapid and flexible prototyping and management of BSN applications.

Due to the distributed nature of sensor networks, which allow connection among several smart objects scattered over large areas, modern distributed computing paradigms have been adopted. In particular, **Agent-based** techniques have been adopted to support cooperation among smart objects and to gather data from interconnected sensors [8, 13]. In order to analyze issues and bottlenecks at communication level within sensor networks, agent based models have been proposed in literature: such models can be used to simulate the network activity and help on designing the most correct interconnections [11]. A meta-model based approach has been proposed in [10], where models defined at different levels of granularity and abstraction, are used as a basis to develop a concrete agent-based solution.

Cloud Computing has been also exploited to provide a robust and scalable infrastructure to sensors networks, in particular for data collection and elaboration. The work presented in [12] provides a clear approach, based on combined Cloud and Agent-based technologies, for the management of highly scalable networks of smart sensors.

The current state of IoT infrastructure does not allow direct access to the information stored and elaborated by the devices, and the only way to access the data is through RESTful API provided by the specific vendors: this particular feature is quite critical as it leaves little space to provide interconnectivity.

The specific applications supplied with the smart devices are the ones that communicate directly with the physical device and an end user or developer can access the data solely through the application itself or better trough the APIs, where provided; even the ways in which the vendor application communicate with the device cannot be known.

Even though the applications' interfaces provided by each vendor are easily accessible, every interfaces has its own representation and specific structure, not standard of course, and in most cases not machine-readable. The work presented in [1] describes an approach, based on opportunistic gateways, to solve interoperability issues the IoT. Interoperability among IoT devices is also the subject of research projects, such as the H2020 funded project **Inter-IoT** [15] that proposes a multi-layered approach in which different IoT devices, networks, platforms, services and applications are integrated to allow a global continuum of data, infrastructures and services. The project has been funded within the **IoT European Platforms Initiative**, which addresses seven projects focusing on different aspects of the IoT ecosystem [16]. The data heterogeneity that characterizes both the APIs' structure and their natural language descriptions, has become a real obstacle that complicates the task of developing connectors that could be adopted by different sensors and moreover they are hard to maintain and update. Recently many framework or software known as "aggregators" have been developed and designed, their main purpose is to provide an unified interface to access to the various provider dependent sensors or restful services in a completely transparent way.

Embed is a commercial aggregator that enables developers to easily embed content from third party content providers like YouTube, Vine, Flickr and many more. Embed follows the **oEmbed** specification. oEmbed is a format for allowing an embedded representation of a URL on third party sites. The simple API allows a website to display embedded content (such as photos or videos) when a user posts a link to that resource, without having to parse the resource directly.

Another example is SocIoS API framework [18]: currently the most popular social networks and media, such as Twitter, Facebook and YouTube expose all or part of their functionality through open RESTful APIs through which every user or third party application can gain access to their content and operations. Despite the similarities in notions and basic functionality, data representation in social networks is highly heterogeneous. In addition to that, each social network offers its own API and due to the lack of a non commercial tool for accessing multiple APIs from a single API, a user looking to combine data from two or more social networks will have to invoke all the APIs and transform the data in a common format before processing them. The SocIoS framework aims to address the above mentioned challenges. It is a software stack that operates on top of Social Networking Sites (SNS). SocIoS provides an abstraction layer for combining data and functionality from a multitude of underlying social media platforms as well as a set of analytical tools for leveraging that functionality. At the core of the SocIoS project, lies the SocIoS API. It constitutes a single access point for a number of popular social networks exposing operations that encapsulate their functionality.

3 Analysis of Past and Current API Representations

As also reported in Sect. 2, there are several efforts towards the definition of an efficient and shareable formalism to describe IoT online services. However, as most of IoT services are nowadays callable via RESTful interfaces, the main research line consists in defining new formalisms for the representation of such interfaces, with a focus on sensors and devices. As a consequence, formalisms for API description have been proposed, also to answer the need to a common representation for sensors' interfaces. In this section, we are going to describe the main actors in the API description scenery and to compare them to each other and to the WSDL standard, widely used in the past for web services definition.

3.1 Web Application Description Language (WADL)

The **Web Application Description Language** (WADL) [20] is a machine readable formalism for the description of HTTP based web-services, based on XML. While Sun Microsystems submitted it to the World Wide Web Consortium (W3C) in 2009, there is still no plan to actually standardize it [19] In respect to WSDL documentation, WADL description are lighter and more straightforward, in order to better adapt to modern web-services interfaces, which generally employ REST. The structure of a WADL document is indeed very simple:

- The core element of the description is represented by the **Resources** tag, which exposes a **base** property to point out the base address of the services.

- The Resources tag contains a collection of **Resource**, each representing the single services which can be accessed from the base address. In particular, the name of the service is identified by the **path** attribute.
- Within a resource, methods are defined via the **Method** tag, exposing a **name** attribute (to state the type of HTTP request issued) and an **id** with the exact name of the called service.
- Each Method describes a **Request** and a **Response**. Both of them contain parameters (**param** tag), with at least a name and a type, but which can also expose information on the obligatoriness or cardinality of the parameter. Furthermore, if the parameter can only be chosen from a predefined set of values, the **option** tag can be used to enumerate them.

The hierarchical model used to obtain the WADL description of the web service can be rapidly parsed to obtain the information required to build the service call. Also, the entire document is much smaller and simpler than a WSDL counterpart. Consider the example provided in listing 1 (published by the W3C itself), describing a search service exposed by Yahoo: it counts less than fifty lines of code (header included), and a human reader could easily understand it. Obviously, being simpler than WSDL, WADL is also less flexible and offers less functionalities. Table 1 provides a comparison between the two languages.

```
<?xml version="1.0"?>
<application xmlns:xsi="http://www.w3.org/2001/XMLSchema-instance"
  xsi:schemaLocation="http://wadl.dev.java.net/2009/02 wadl.xsd"
  xmlns:tns="urn:yahoo:yn"
  xmlns:xsd="http://www.w3.org/2001/XMLSchema"
  xmlns:yn="urn:yahoo:yn"
  xmlns:ya="urn:yahoo:api"
  xmlns="http://wadl.dev.java.net/2009/02">
   <grammars>
      <include
         href="NewsSearchResponse.xsd"/>
      <include
         href="Error.xsd"/>
    </grammars>
     <resources base="http://api.search.yahoo.com/NewsSearchService/
        V1/">
      <resource path="newsSearch">
        <method name="GET" id="search">
          <request>
            <param name="appid" type="xsd:string"
              style="query" required="true"/>
            <param name="query" type="xsd:string"
              style="query" required="true"/>
            <param name="type" style="query" default="all">
              <option value="all"/>
              <option value="any"/>
              <option value="phrase"/>
            </param>
            <param name="results" style="query" type="xsd:int"
                default="10"/>
            <param name="start" style="query" type="xsd:int" default
                ="1"/>
            <param name="sort" style="query" default="rank">
              <option value="rank"/>
              <option value="date"/>
            </param>
            <param name="language" style="query" type="xsd:string"/>
```

```
            </request>
            <response status="200">
              <representation mediaType="application/xml"
                element="yn:ResultSet"/>
            </response>
            <response status="400">
              <representation mediaType="application/xml"
                element="ya:Error"/>
            </response>
          </method>
        </resource>
      </resources>
    </application>
```

Listing 1 An example of a WADL description for the Yahoo News Search application

3.2 *RESTful API Modeling Language (RAML)*

The **RESTful API Modeling Language** (RAML) [24] is a vendor-neutral and open specification language built on YAML [3] and JSON for describing RESTful APIs. First proposed in 2014 by the **RAML Workgroup**, it has been taken in consideration by the OpenApi consortium, together with other API specification languages, for standardization. Similarly to other languages for the description or REST and REST-like interfaces, RAML specification tries to be as simple as possible and to offer a lightweight description of web-services. However, it also supports the definition of patterns and data type inheritance, in a fashion very similar to object-oriented languages. The structure of a RAML document is composed of **nodes**, each describing a peculiar element of the API interface:

- **Title** and **Description** respectively contain a string label which identifies the service and a human-friendly description of it. The Description, in particular, should provide directions for the use of the service.
- The **baseURI** and **baseURIParameters** nodes represent fixed elements of every service call. In particular, baseURI defines the base for URIs of all resources. Often used as the base of the URL of each resource. Instead, baseURIParameters defines all the named parameters which are used in the base URI.

Table 1 A comparison between WSDL and WADL

WSDL	WADL
Complex design	Simple design
Difficult to read from a human point of view.	Easy to read and implement
Supports all HTTP verbs and several other protocols (e.g. SMTP)	Only supports HTTP
Accepts XML parameters only	Parameters types are not limited to XML
W3C recommendation	Not standardised
Authorization mechanisms available	No authorization

- The **types** node can be used to define new data types (named types), which can be also declared inline (unnamed types) or in external libraries. Both named and unnamed types need to declare a **schema** or **type** node (they are mutually exclusive and their use depends on the RAML version used), which can refer to one of the built-in data types provided by RAML, or to another named type defined elsewhere.
  - **Object types** are a particular built-in type that can declare additional **properties**, so they can be used to define complex parameters
- A **Resource** is identified by its relative (to the baseURI node) URI, and it is identified with a slash. It can be defined at root level (top-level resource) or nested within another resource. Its three main components are represented by:
  - a **Method** node containing the list of parameters needed to call the specific service described by the resource (via the mutually exclusive **queryParameters** or **queryString** nodes), the description of the **header** and of the **response** expected from the method call.
  - A **Resource Type** that the current resource inherits
  - A list of **Traits** that apply to all the methods described by the resource, which can be overridden by the specific method.
- **Resource Types** and **Traits** are at the core of the inheritance schema implemented by RAML, and contribute to the re-use of code and to its consistency and maintenance. A Resource Type specifies a generic resource whose methods and properties can be inherited by actual resources definitions. Traits are instead applied to all the methods exposed by resources, and define characteristics that they share.
- Most REST APIs have one or more mechanisms to secure data access, identify requests, and determine access level and data visibility. The **securitySchemas** node describes the security schemas definitions supported by the API.

RAML tends to be a little more complex than WADL, as it also supports inheritance and code re-use. Also, it can be difficult to read for users who have small or no experience at all with YAML, but are instead more experienced in XML-based documentation. This is not a real problem, as after a very short time using it, users can definitely learn how it is structured and use it seamlessly.

Listing 2 presents a RAML description of the Yahoo News Search application. As in the WADL representation, in which responses were defined in external XML documents, here we suppose that external RAML files are available for inclusion (via the **!include** command).

```
/newsSearch:
/{search}
get:
queryParameters:
appid:
displayName: Application ID
type: string
description: The ID of the Application
required: true
query:
displayName: Query
type: string
description: The query to be searched
required: true
type:
displayName: Type
type: string
description: The Type of query
required: true
default: "all"
enum: ["all","any","phrase"]
results:
displayName: Results
type: int
description: The number of results to return
required: false
default: 10
sort:
displayName: Sort
type: string
description: The ordering criterion for results
required: false
default: "rank"
enum: ["rank","date"]
start:
displayName: Start
type: int
description: The index of the result to display
required: false
default: 1
responses:
200:
body:
application/xml:
type: !include ResultSet.yaml
400:
body:
application/xml:
type: !include Error.yaml
```

Listing 2: A RAML representation of the Yahoo News Search Application

Table 2 provides a three-way comparison among WSDL, WADL and RAML.

Table 2 Comparison between WSDL, WADL and RAML

	Design	Reading and implementation	Protocols	Parameters	Standard	Authentication	Reuse support
WSDL	Complex	Difficult	HTTP and others	XML	W3C	Yes	Data types extension
WADL	Simple	Easy	HTTP	XML, JSON	No	No	Data types extension
RAML	Medium	Requires knowledge of YAML	HTTP	XML, JSON, YAML built-in	No	Yes (Schemas)	Inheritance

3.3 API Blueprint

API Blueprint [4] is a documentation-oriented web API description language, built upon the **Markdown** syntax [17], which is a plain-text syntax for formatting documents that can be immediately translated in HTML pages via the Markdown tool. An API Blueprint document is a plain text Markdown document describing a Web API, structured into logical **sections**. Such section have a specific location within the documentation, can be nested and, while completely optional, if present they must follow the Blueprint formalism. The language reserves some keywords to identify the section types, so that they cannot appear as the identifying names of a section. As an instance, HTTP verbs (GET, POST, DELETE...) are keywords in Blueprint. Therefore, a section is composed of:

- A **Keyword** with the **Identifier** of the section (its unique name);
- A section's description, which is any arbitrary Markdown-formatted content following the section definition. It can contain reserved keywords, as it is treated as a comment to the section.
- Content specific for the described section
- Nested sections

Also, it is possible to distinguish between two main categories of sections: **Abstract** sections need to be extended as they cannot be used directly; **Section Basics** are instead directly usable to build sections. Among Abstract sections, the language defines:

- A **Named Section** represents the base of all other API Blueprint sections, as it is composed by an identifier, a description and nested sections, which can be alternatively substituted by specific formatted content;
- An **Asset Section** represent the base for all atomic data in Blueprint, as it is described by a pre-formatted code block.
- A **Payload Section** represents the payload transferred as part of an HTTP request or response.

Section Basics define the main building blocks of the API Blueprint documentation, and they are represented by:

- The **Metadata Section** which is composed of key-value pairs separated by a semicolon. They provide metadata annotations which are tool specific.
- The **API name & overview section** is the first header in a Blueprint document, as it presents the name and description of the API. It inherits from Named Section.
- The **Resource group** section, identified by the **Group** keyword, represent a group of resources, thus it may include one or more Resource Sections.
- A **Resource Section** represents an API resource, specified by its URI. The formalism allows for four different kinds of Resource section instantiations:

  1. A simple URI template;
  2. An identifier followed by the URI template in square brackets;
  3. An HTTP request method followed by a URI template;
  4. An identifier followed by An HTTP request method and a URI template, in square brackets.

  In the last two cases, the remainder of the Resource section follows the specifics of the Action Section. A Resource section must contain at least one Action Section, an it can contain additional optional sections, such as the Attributes Section.
- The **Attribute Section** describe attributes of a resource, an action or a payload. Named attributes can be referenced by other sections. They are defined via the **Attributes** keyword, followed by an optional MSON (Markdown Syntax for Object Notation) Type. If omitted, the attribute is considered as an object and defines a structured data type containing more attributes.
- The **Action Section** can be introduced by either an HTTP request method, an action name followed by an HTTP request method enclosed in square brackets, or by an action name followed by an HTTP request method and URI template enclosed in square brackets. It is always nested within a Resource Section and it provides the definition of at least one HTTP transaction as performed with the parent resource section. One and only one Parameter section can be defined within an Action, while optional Attributes defined in the action are included as input of the nested Request sections. Multiple Request and Response Sections may be nested in the Action section.
- A **Parameter** Section describes the URI parameters in a Markdown list item. It defines the parameter name, **default** value, **type** and list of possible values the parameter can assume (using the **Members** optional keyword). Each parameter can be **required** or **optional**

The language is very simple to read, even for non experts, and the use of the Markdown syntax surely helps. The specification supports the use of JSON and XML types for HTTP responses and requests, via **Schema Sections**, which describe how JSON and XML data structure should be formatted.

Table 3 Comparison between WSDL, WADL, RAML and Blueprint

	Design	Reading and implementation	Protocols	Parameters	Standard	Authentication	Reuse support
WSDL	Complex	Difficult	HTTP and others	XML	W3C	Yes	Data types extension
WADL	Simple	Easy	HTTP	XML, JSON	No	No	Data types extension
RAML	Medium	Requires knowledge of YAML	HTTP	XML, JSON, YAML built-in	No	Yes (Schemas)	Inheritance
Blueprint	Simple	Easy—Requires knowledge of markdown syntax	HTTP	MSON, JSON, XML	No	No	Named Attributes can be reused in different sections

Listing 3 provides the Blueprint version of the Yahoo News Search API we have used in previous section, while Table 3 provides a comparison among all the formalisms described so far.

```
RMAT: 1A
ST: http://api.search.yahoo.com/NewsSearchService/V1/
API
NewsSearch [/newsSearch{?appid,query,type,results,start,sort,language}]
 search [GET]
Parameters
 appid (string, required)
 query (string, required)
 type (enum[string], optional)
+ Default: all
+ Members
        + 'all'
        + 'any'
        + 'phrase'

 results (number, optional) -
+ Default: 10
 start (number, optional) -
+ Default: 1
+ sort (enum[string], optional)
        + Default: rank
        + Members
                + 'rank'
                + 'date'
+ language (string, optional)
Response 200
Attributes (string)
Response 400
```

Listing 3: A Blueprint representation of the Yahoo News Search Application

4 Swagger (Now as OpenAPI)

Swagger [26] is the former name of both an API specification language and of a framework implementation based on it, which aims at providing a standard representation for APIs, together with a both human and machine readable documentation. Originally developed for Wordnik [7] to support the Wordnik Developer and its underlying API, it was acquired by SmartBear which, in 2015, founded the OpenAPI Initiative, under the sponsorship of the Linux Foundation. SmartBear donated the Swagger specification to the new group, which renamed it as the **OpenAPI Specification**. RAML and API Blueprint are also under consideration by the group.

One of the strong points of the documentation is the ample use of JSON: files describing the RESTful API in accordance with the Swagger specification are represented as JSON objects and conform to the JSON standards. Being YAML a superset of JSON, YAML parsers are able to understand Swagger documents. Also, the adoption of the JSON standard does not limit the type of attributes which can be defined in the API interaction.

The Swagger specification is based on nested objects. The root document, called **Resource Listing** is basically a collection of **Resource Objects**, each providing the **path** to a resource reachable through the API. The Resource Listing also provides information on the Swagger and API versions, additional information (via the **Info Object**), and supports Authorization (**Authorization Object**).

The resources declared via the Resource Listing are described by a correspondent **API Declaration**. An API Declaration provides information about an API exposed on a resource. In particular, it exposes the root URL serving the API (**basePath**) and the relative path to the resource (**resourcePath**). Authorization schemes can be defined, as for Resource Listings, via Authorizations Objects.

The core of an API Declaration is represented by the **API Object**, which describe one or more possible operations possible on a single path. Each API Object provides the **path** to the operation, its description and list of **Operation Objects**.

An Operation Object describes a single operation on a path. It contains the declaration of the called **method** (an HTTP verb), a unique id to identify the operation (**nickname**), and a list of parameters, expressed via **Parameter Objects**. Response Messages are described via **Response Message Objects** instead, which contains a response code and a message.

Parameter Objects describe a single parameter to be sent in an operation. Each object declares a type, which can be chosen among the values "path", "query", "body", "header" and "form", and a name which strictly depends on the type and on the path to the operation.

Listing 4 reports a Swagger-based representation of the Yahoo News Search Application, while Table 4 extends previous Table 3 by adding Swagger to the comparison.

```
{
  "swagger": "2.0",
  "info": {
    "version": "1.0",
    "title": "API",
    "license": {
      "name": "MIT",
      "url": "http://github.com/gruntjs/grunt/blob/master/LICENSE-MIT"
    }
  },
  "host": "api.search.yahoo.com",
  "basePath": "/NewsSearchService/V1/",
  "schemes": [
    "http"
  ],
  "paths": {
    "/newsSearch": {
      "get": {
        "operationId": "search",
        "produces": [
          "application/json"
        ],
        "parameters": [
          { "name": "appid", "in": "query", "required": true, "type": "string",},
          { "name": "query", "in": "query", "required": true, "type": "string",},
          { "name": "type", "in": "query", "required": false,
            "enum": [   "all",    "any",    "phrase" ], "default": "all", "type": "string",},
          { "name": "results", "in": "query", "required": false, "default": 10.0,
            "type": "integer", "format": "int32",},
          { "name": "start", "in": "query", "required": false, "default": 1.0,
            "type": "integer", "format": "int32",},
          { "name": "sort", "in": "query", "required": false, "enum": [    "rank",    "date" ],
            "default": "rank", "type": "string",},
          { "name": "language", "in": "query", "required": false, "type": "string",}
        ],
        "responses": {
          "200": {"schema": {    "type": "string" }
          },
          "400": {"schema": {}
          }
        }
      }
    }
  },
  "definitions": {
    "Type": {
      "title": "type",
      "type": "string",
      "enum": [
        "all",
        "any",
        "phrase"
      ]
    },
    "Sort": {
      "title": "sort",
      "type": "string",
      "enum": [
        "rank",
        "date"
      ]
    }
  }
}
```


Listing 4: A Swagger representation of the Yahoo News Search Application

Table 4 Comparison between WSDL, WADL, RAML, blueprint and swagger

	Design	Reading and implementation	Protocols	Parameters	Standard	Authentication	Reuse support
WSDL	Complex	Difficult	HTTP and others	XML	W3C	Yes	Data types extension
WADL	Simple	Easy	HTTP	XML, JSON	No	No	Data types extension
RAML	Medium	Requires knowledge of YAML	HTTP	XML, JSON, YAML built-in	No	Yes (Schemas)	Inheritance
Blueprint	Simple	Easy—requires knowledge of markdown syntax	HTTP	MSON, JSON, XML	No	No	Named attributes can be reused in different sections
Swagger	Medium	Medium–JSON is not always immediately readable	HTTP	XML,JSON, XML	No	Yes	Objects can be reused in several locations

5 Analysis of Existing APIs

In order to obtain a coherent, machine readable representation of existing sensors' APIs, described via any of the formalisms already presented in Sect. 3, it is necessary to actually discover and analyse such APIs. Several approaches are possible, but their feasibility and efficiency need to be taken in serious consideration.

Obviously, the most effective approach would be to analyse an already existing machine readable and standardized representation of the published APIs, and then to convert it to one of the preferred formalisms. This is, as an instance, the methodology applied in the FP7 mOSAIc project [23], in which an API parser was built to retrieve information on the existing API calls and to implement a **Dynamic Discovery Service** for Cloud platforms [6]. However, the parsing tool was only able to analyse already formatted and standardized representations of APIs, either provided in a semantic-based language such as OWL [21] or OWL-S [5], or in the very common (when the mOSAIc project was running) WSDL format. As of today, such an approach would be extremely limited and limiting: first of all, semantic descriptions of APIs are not common at all, even less in the IoT domain; second, even WSDL description of APIs and services are slowly but steadily disappearing. WSDL descriptions of web services have been deprecated almost everywhere: just consider that Amazon, among the pioneers and main stakeholders of the Cloud Computing era, have deprecated the WSDL descriptions of their historical **Amazon Web Services** in favour of new RESTful interfaces. This does not mean that WSDL

```
-<xs:complexType name="ItemSearchRequest">
  -<xs:sequence>
     <xs:element name="Actor" type="xs:string" minOccurs="0"/>
     <xs:element name="Artist" type="xs:string" minOccurs="0"/>
   -<xs:element name="Availability" minOccurs="0">
     -<xs:simpleType>
       -<xs:restriction base="xs:string">
           <xs:enumeration value="Available"/>
         </xs:restriction>
       </xs:simpleType>
     </xs:element>
     <xs:element ref="tns:AudienceRating" minOccurs="0" maxOccurs="unbounded"/>
     <xs:element name="Author" type="xs:string" minOccurs="0"/>
     <xs:element name="Brand" type="xs:string" minOccurs="0"/>
     <xs:element name="BrowseNode" type="xs:string" minOccurs="0"/>
     <xs:element name="Composer" type="xs:string" minOccurs="0"/>
     <xs:element ref="tns:Condition" minOccurs="0"/>
     <xs:element name="Conductor" type="xs:string" minOccurs="0"/>
     <xs:element name="Director" type="xs:string" minOccurs="0"/>
     <xs:element name="ItemPage" type="xs:positiveInteger" minOccurs="0"/>
     <xs:element name="Keywords" type="xs:string" minOccurs="0"/>
     <xs:element name="Manufacturer" type="xs:string" minOccurs="0"/>
     <xs:element name="MaximumPrice" type="xs:nonNegativeInteger" minOccurs="0"/>
     <xs:element name="MerchantId" type="xs:string" minOccurs="0"/>
     <xs:element name="MinimumPrice" type="xs:nonNegativeInteger" minOccurs="0"/>
     <xs:element name="MinPercentageOff" type="xs:nonNegativeInteger" minOccurs="0"/>
     <xs:element name="MusicLabel" type="xs:string" minOccurs="0"/>
     <xs:element name="Orchestra" type="xs:string" minOccurs="0"/>
     <xs:element name="Power" type="xs:string" minOccurs="0"/>
     <xs:element name="Publisher" type="xs:string" minOccurs="0"/>
     <xs:element name="RelatedItemPage" type="tns:positiveIntegerOrAll" minOccurs="0"/>
     <xs:element name="RelationshipType" type="xs:string" minOccurs="0" maxOccurs="unbounded"/>
     <xs:element name="ResponseGroup" type="xs:string" minOccurs="0" maxOccurs="unbounded"/>
     <xs:element name="SearchIndex" type="xs:string" minOccurs="0"/>
     <xs:element name="Sort" type="xs:string" minOccurs="0"/>
     <xs:element name="Title" type="xs:string" minOccurs="0"/>
     <xs:element name="ReleaseDate" type="xs:string" minOccurs="0"/>
     <xs:element name="IncludeReviewsSummary" type="xs:string" minOccurs="0"/>
     <xs:element name="TruncateReviewsAt" type="xs:nonNegativeInteger" minOccurs="0"/>
   </xs:sequence>
```

Fig. 2 Example of WSDL document (excerpt)

is going to disappear, since its support to services composition and discovery cannot be provided by current RESTful interfaces as they are. However, while being extremely useful to identify APIs and automatically call web services, WSDL has proved to be too slow and computational consuming, since clients need to analyse long and complex XML documents, and then format the request message following the very precise structure described in them. The same happens to servers each time they reply to a received request. Just consider the very small (in comparison to the whole document) WSDL excerpt reported in Fig. 2, describing just one of the complex parameters that can be exchanged via SOAP in and API request made to a former Amazon web service. Considering the high volume of traffic that servers need to manage continuously and the very limited computational capabilities owned by mobile devices, which are becoming the main access points to on-line services as of today, the shift towards a lighter and faster way to define APIs was inevitable. This is event truer for smart devices and sensors that not only have a very limited computational power, but also need to save as much energy as possible due to the very limited batteries they own.

In an ideal world, RESTful interfaces would be accurately described and/or a machine readable definition thereof would be publicly available. In such a utopistic

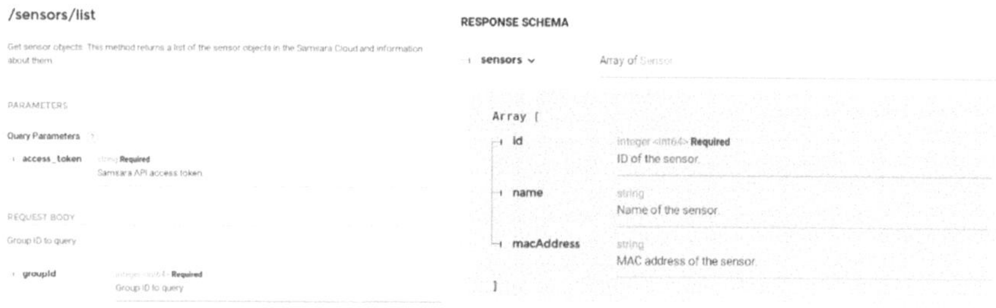

Fig. 3 Description of request and response of Samsara API

scenario, it would be possible to automatically analyse the Natural Language or machine readable descriptions of RESTful services and automatically produce the service calls to exploit a specific functionality, and then read the answer appropriately. Even considering the very different semantics used in the available interfaces, it would be possible to produce the "wrappers/adapters" needed to translate the inputs and outputs for each specific vendor endpoint. Developers have several tools at their disposal to automatically produce machine readable descriptions of their RESTful services: **API Workbench** for RAML [2] is just an example. Nevertheless, while we have the technologies and the standards (or at least proposals of standards) to provide such descriptions, in a real world scenario the situation is far from ideal.

In some, very rare cases, textual descriptions of the input and output parameters of the POST and GET calls of REST interfaces are present. As an instance consider the description provided at [25] and reported in Fig. 3 for commodity. In the reported case, Request and Response are inserted into a HTML table, with a description of the parameters in natural language. In such a situation very simple analysis of the descriptions, offered in structured and static HTML pages, can be driven and the necessary information can be retrieved. This is, however, a very simplistic and not generalizable approach, for two obvious reasons:

1. If a crawler was built to analyse such a page, it would be programmed ad-hoc for its HTML structure, and it would be absolutely useless for a different one. Knowing the HTML structure of each documentation page of different APIs represents an almost unsatisfiable requirement. Furthermore, a different crawler would have to be developed.
2. Having a structured description of the APIs is still rare as it represents a very small minority of the actual cases.

Indeed, most of the times the pages, even when showing a definite structure, are dynamic: the descriptions integrated into web pages are produced at run time through scripting code client/server side (e.g. JavaScript/PHP). Such a code is not accessible via a simple page crawling, thus making it impossible to analyse the descriptions automatically. This is the case of the API provided by **Yahoo Weather** [27]. The sample Request and Response code of the API reported on this page, whose screen-shot has been reported in Fig. 4 is indeed automatically generated by selecting the desired

Fig. 4 Page describing Yahoo weather API

options on the very same page. This means that, except from the page template, no other information can be retrieved from the structural analysis of the HTML source. It would be theoretically possible to analyse each possible input contemplated by the form used to build the call to the server side script, in order to retrieve all possible outputs. But, that would require a an additional, not trivial effort which could also be rendered useless if there were free forms, with no options to actually select.

Another, unfortunately, very common scenario is represented by descriptions built up of undocumented JSON string examples, used as input or output of the calls, where the name of the called service is just one of the passed parameters. That is the case, for example, of the already cited REST API described in [27]. In such cases, Natural Language Processing techniques can be used to analyse the on-line documentation and determine either where the parameters description is, or in which point of the web-pages the JSON example is reported. Then, it is possible to analyse both the parameters and the JSON string, by using string-based matching techniques, making it possible to understand the meaning of each parameter and of the function call in general. Again, the success of such a technique depends from how the documentation page has been produced, and if the displayed JSON is actually available for a web crawler/parser.

6 Authorization and Authentication Issues in API Aggregation

The paradigm of offering a service through a high available restful interface facilitates the software artefacts integration and interoperability: to perform an action using a web service's API, you need to select a calling convention, send a request to its endpoint specifying a method and some arguments, and will receive a formatted response. In this chapter we have even pointed out that the further step of creating an

APIs aggregator may even more enable the development of more efficient software, however when it comes to this solution there is a critical aspect that must be addressed in this field that is the *authorization and authentication mechanism*. In respect to WSDL representation, in which there is a distinct part of the structure dedicated to the authentication and authorization process, the current restful API representations do not provide it, the reason for this feature can be found in the way a developer accesses to the web services.

Many if not every single API method requires the user to be logged in to access the services. At present there is only one way to accomplish this: users should be authenticated using the specific application's Authentication API, through this end point, a developer can retrieve an *application API Key*, that is essentially an *authentication token*, but the developer must send a request to the end point specifying an username and a password. Once these *tokens* are retrieved they could be used in all the following requests. An easy solution for using the API aggregator once the various tokens are retrieved by the user, is to store them all so that the API aggregator should be able to automatically call different vendor's API. However even though the tokens are stored, some of them have a limited time frame, so tokens could even change after each request.

Another way to support the use of the API aggregator and to let it handle every step of the API request and response process it could be to allow the aggregator to automatically create a send request for every services' Authentication API and retrieve every single *API Key* and use them accordingly to the service that has to be called. For the API aggregator to create the custom request and to obtain the token the user has to provide the various usernames and passwords, but this specific step may arise various security issues:

- The user credentials must be sent in a secure and reliable way.
- The API aggregator could store the users' credentials, but this solution assumes that a number of features must be granted, such as security, reliability and confidentiality.
- The API aggregator may not save the users' credentials requesting them on demand according to the service to be called; however the aggregator must ensure that in no way possible the credentials are stored, not even temporarily, and that they cannot be retrieved by third party applications and machines.

7 Conclusion

In this chapter we have tried to determine and summarize the main technologies available, as of today, to represent and share APIs, which are or can be exploited to represent IoT interfaces. We have seen that, despite the existence of several candidates for a correct and shareable representation of such IoT APIs, in a real world scenario they are still hardly used, as the most relevant vendors tend to just propose non standardized description of their RESTful interfaces. Also, we have stressed the fact that an automatic analysis of current non formal description would be unfeasible.

At the end of the day, we currently have necessary technologies to actually develop an integrate IoT framework, thanks to the several formalisms which are available for APIs definition. However, the lack of extensive use of such formalisms and the fact that they are not standardized yet (except for WADL), surely represent an obstacle to outcome.

References

1. Aloi, Gianluca, Giuseppe Caliciuri, Giancarlo Fortino, Raffaele Gravina, P. Pace, Wilma Russo, and Claudio Savaglio. 2017. Enabling iot interoperability through opportunistic smartphone-based mobile gateways. *Journal of Network and Computer Applications* 81: 74–84.
2. API workbench. http://apiworkbench.com. Accessed 8 Feb 2017.
3. Ben-Kiki, Oren, Clark Evans, and Brian Ingerson. 2005. Yaml ain't markup language (yaml) version 1.1. *yaml. org, Tech. Rep.*
4. Blueprint, A. P. I. Format 1A revision 8. https://github.com/apiaryio/api-blueprint/blob/master/API%20Blueprint%20Specification.md, 05–22.
5. Burstein, Mark, Hobbs Jerry, Lassila Ora, Mcdermott Drew, Mcilraith Sheila, Narayanan Srini, Paolucci Massimo, Parsia Bijan, Payne Terry, Sirin Evren, Srinivasan Naveen, and Sycara Katia. 2004. OWL-s: Semantic markup for web services. http://www.w3.org/Submission/2004/SUBM-OWL-S-20041122/.
6. Cretella, Giuseppina, and Beniamino Di Martino. 2013. Semantic and matchmaking technologies for discovering, mapping and aligning cloud providers's services. In *Proceedings of the 15th international conference on information integration and web-based applications and services (iiWAS2013)*, 380–384.
7. Davidson, Sara. 2013. Wordnik. *The Charleston Advisor* 15(2): 54–58.
8. Fortino, Giancarlo, Antonio Guerrieri, and Wilma Russo. 2012. Agent-oriented smart objects development. In *Proceedings of the 2012 IEEE 16th international conference on computer supported cooperative work in design (CSCWD)*, 907–912.
9. Fortino, Giancarlo, Roberta Giannantonio, Raffaele Gravina, Philip Kuryloski, and Roozbeh Jafari. 2013. Enabling effective programming and flexible management of efficient body sensor network applications. *IEEE Transactions on Human-Machine Systems* 43(1): 115–133.
10. Fortino, G., A. Guerrieri, W. Russo, and C. Savaglio. Towards a development methodology for smart object-oriented iot systems: A metamodel approach. In *2015 IEEE international conference on systems, man, and cybernetics*, 1297–1302, Oct 2015.
11. Fortino, G., W. Russo, and C. Savaglio. Agent-oriented modeling and simulation of iot networks. In *2016 federated conference on computer science and information systems (FedCSIS)*, 1449–1452, Sept 2016.
12. Fortino, G. A. Guerrieri, W. Russo, and C. Savaglio. Integration of agent-based and cloud computing for the smart objects-oriented iot. In *Proceedings of the 2014 IEEE 18th international conference on computer supported cooperative work in design (CSCWD)*, 493–498, May 2014.
13. Giancarlo Fortino, Antonio Guerrieri, Michelangelo Lacopo, Matteo Lucia, and Wilma Russo. 2013. *An agent-based middleware for cooperating smart objects*, 387–398. Berlin Heidelberg: Springer.
14. Gravina, Raffaele, Parastoo Alinia, Hassan Ghasemzadeh, and Giancarlo Fortino. 2017. Multi-sensor fusion in body sensor networks: State-of-the-art and research challenges. *Information Fusion* 35: 68–80.
15. Inter-iot. http://www.interiot.eu. Accessed July 2017.
16. Iot european project initiative. http://iot-epi.eu/projects. Accessed July 2017.

17. John Gruber. Markdown: Syntax. http://daringfireball.net/projects/markdown/syntax. Accessed 24 June 2012.
18. Kardara, Magdalini, Vasilis Kalogirou, Athanasios Papaoikonomou, Theodora Varvarigou, and Konstantinos Tserpes. 2014. Socios api: A data aggregator for accessing user generated content from online social networks. In *International conference on web information systems engineering*, 93–104. Springer.
19. Lafon, Y. 2009. Team comment on the web application description language submission. http://www.w3.org/Submission/2009/03/Comment. Accessed August 2011.
20. Marc J Hadley. Web application description language (wadl). 2006.
21. McGuinness, Deborah L., Frank Van Harmelen, et al. 2004. *Owl web ontology language overview*. 10(10).
22. Miorandi, Daniele, Sabrina Sicari, Francesco De Pellegrini, and Imrich Chlamtac. 2012. Internet of things: Vision, applications and research challenges. *Ad Hoc Networks* 10(7): 1497–1516.
23. Petcu, Dana, Beniamino Di Martino, Salvatore Venticinque, Massimiliano Rak, Tamás Máhr, Gorka Esnal Lopez, Fabrice Brito, Roberto Cossu, Miha Stopar, Svatopluk Šperka, and Vlado Stankovski. Experiences in building a mosaic of clouds. *Journal of Cloud Computing: Advances, Systems and Applications* 2(1): 12.
24. RAML Workgroup.2015. Raml-restful api modeling language. http://raml.org/ 2015. Accessed 10 Feb 2017.
25. Samsara web-Site. https://www.samsara.com/api. Accessed 8 Feb 2017.
26. Swagger Team. 2014. Swagger restful api documentation specification 1.2. Technical report, Technical report, Wordnik. https://github.com/wordnik/swagger-spec/blob/master/versions/1.2.md.
27. Yahoo weather API. https://developer.yahoo.com/weather/. Accessed on 8 Feb 2017.

Energy Harvesting in Internet of Things

Cheuk-Wang Yau, Tyrone Tai-On Kwok, Chi-Un Lei and Yu-Kwong Kwok

Abstract Powering billions of connected devices has been recognized as one of the biggest hurdles in the development of Internet of Things (IoT). With such a volume of tiny and ubiquitous smart physical objects in this new Internet paradigm, power cables or sizable battery packs are no longer a viable option to bring them online for years and decades. Energy harvesting, which enables devices to be self-sustaining, has been deemed a prominent solution to these constraints. This chapter provides a comprehensive review of IoT devices, from their roles and responsibilities, to the challenges of operating them autonomously in heterogeneous environments. The concepts, principles and design considerations for energy harvesting are introduced to aid researchers and practitioners to incorporate this key technology into their next applications.

1 The Internet of Things Landscape

The rapid growth of the Internet during the past decades gradually transformed the way humans exchange information. From websites and emails to various forms of social media, the proliferation of the Internet has accelerated the migration from face-to-face and paper-based interaction to electronic communication via computing devices, such as personal computers and smartphones. Despite the evolution in the form of interaction, the majority of popular Internet applications nowadays concentrate on digitizing human-to-human communication, as well as reducing the communication overhead and latency. This paradigm can be interpreted as the *human-driven Internet*. On the contrary, the trend of Internet of Everything (IoE) involves a paradigm shift from the human-driven Internet to the *data-driven Internet*, and has been gaining momentum worldwide [39, 78, 118]. According to Evans [33], the vision of IoE refers to the utilization of environmental *data* collected by intelligent *things*

C.-W. Yau (✉) · T. T.-O. Kwok · C.-U. Lei · Y.-K. Kwok
Department of Electrical and Electronic Engineering, The University of Hong Kong, Pokfulam, Hong Kong
e-mail: cwyau@eee.hku.hk

B. Di Martino et al. (eds.), *Internet of Everything*, Internet of Things,
https://doi.org/10.1007/978-981-10-5861-5_3

to improve various *processes* of *people*, through leveraging the established and the latest Internet infrastructure and technologies.

The concept of *things* in this vision, also known as Internet of Things (IoT), has been considered to be the vital pillar of IoE among the wide variety of definitions proposed by the academia and the industry [1, 33, 39]. The *things*, which are smart physical objects embedded with computational power, sensors and actuators, serve a major role in this paradigm, thanks to their capability to connect the physical world to the Internet. This chapter addresses the challenges of powering the billions and even trillions of connected *things* in the near future, as well as the opportunities for adopting energy harvesting technologies to tackle such challenges.

1.1 Overview of Internet of Things

The "Internet of Things" originated as a marketing term in 1999 to envision the future of incorporating radio-frequency identification (RFID) technology into the Internet, according to Ashton [5]. With the advance of information technology in various areas, the scope of IoT has vastly expanded and led to numerous debate on its definition since then. Although there is not a unified definition of IoT [9, 39], its general concept can be described with the one proposed by the European Technology Platform on Smart Systems Integration (EPoSS) [29], "a worldwide network of interconnected objects uniquely addressable, based on standard communication protocols." Based on this definition, IoT can be further interpreted from the *Internet* and the *Things* perspectives. In this context, *Internet* can be any public or private computer network based on the standard Internet Protocol (IP), while the *Things* refer to objects that bridge the physical and digital worlds, and possess *Internet* connectivity.

Connecting the *Things* to the *Internet* initiates a shift from the human-driven Internet to the data-driven paradigm. In the human-driven Internet, humans play the primary role in supplying and consuming information from the Internet in most applications, such as websites, emails and social media. Such information, often in the form of text, image, video and audio, is in general supposed to be perceived by humans, and hence to facilitate human-to-human (H2H) communication [65]. The development of the human-driven Internet can be considered as a progression of H2H interaction modes with drastically improved efficiency of long-distance communication, rather than a fundamental transformation of humans' interaction modes with the surroundings. The *Things* may ultimately become the first-class citizens in the data-driven Internet, as they are capable of achieving large-scale automatic environmental sensing, control and machine-to-machine (M2M) communication with their embedded electronics.

In this IoE-inspired paradigm depicted in Fig. 1, the *Things* are responsible for acquiring an unprecedentedly large volume of data from the natural and built environment. The collected data can be analyzed and visualized as meaningful information using Internet services. Such visualized data and insights can assist humans to improve public policies, as well as everyday business and household decisions.

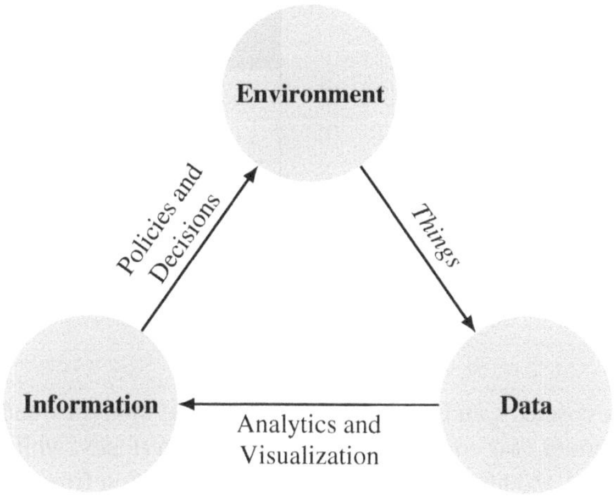

Fig. 1 Illustration of the data-driven Internet paradigm

In addition, computers may utilize the acquired data to optimize the processes autonomously [78]. Eventually, the *Things* may also be able to respond to the environment with their embedded actuators, according to the devised decisions and policies. A feedback loop can hence be formed to enable more effective understanding and efficient management of the surroundings. It is anticipated that with IoT technologies, the manual and often labor-intensive tasks in the loop, such as taking measurements and controlling machines, can be accurately and automatically accomplished by the *Things*, and thus precious human resources can focus on formulating high-level decisions and policies based on data and analytics.

1.2 The Heterogeneous Nature

Heterogeneity marks the most critical characteristic of IoT. As implied by the aforementioned definition and the generalized model for the data-driven Internet, IoT covers a wide spectrum of existing and next-generation applications, ranging from the enterprise level to the consumer grade. IoT applications are designed to facilitate automated and more intelligent processes in different domains, including but not limited to transportation, logistics, health care, emergency services, utilities, agriculture, building and environmental management [9, 39]. Such applications have already been extensively studied by researchers, and common examples include "smart city", "smart building", "smart home" and mobile health systems [54, 78, 117, 118].

In spite of the difference in the nature of IoT applications, the vast majority of them share a common collection of enabling technologies, as depicted in Fig. 2. IP networks, in particular wireless networks that allow *Things* to operate everywhere untethered, serve as the foundation of IoT. On top of IP networks, there are four building blocks for IoT applications, including radio-frequency identification, wire-

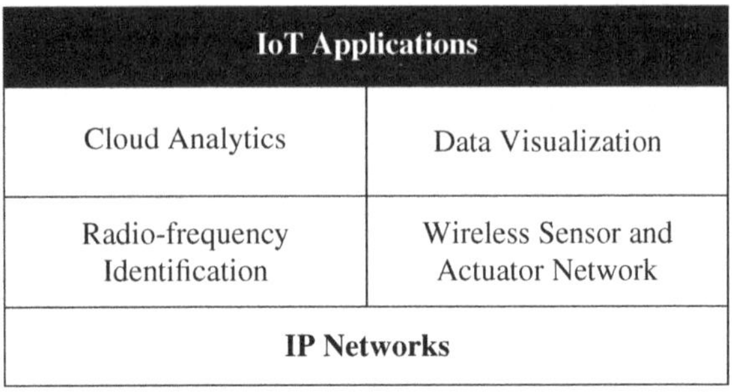

Fig. 2 Enabling technologies of Internet of Things

less sensor and actuator network, cloud analytics and data visualization [39]. The former two construct cyber-physical interfaces, while the latter two compile high-level and human-comprehensible information from raw data.

Radio-frequency identification (RFID) technology allows computers to identify a unique physical object with an attached "tag" [114]. The tag consists of an antenna and small memory (typically a few kilobytes) to store the properties about the object, for example, the identifier and the price of a product in a retail store. When a tagged object is placed in a close proximity to a RFID reader, the tag is activated and its information can hence be retrieved or updated. RFID tags have no or very limited computational power for reading and writing the memory only. This technology has already been extensively used in many industrial applications, from supply chain management to contactless access control and payment cards. Since RFID tags have no capability to sense or interact with the physical world independently, they are regarded as the "passive" *Things* in the IoT paradigm.

The *Things* other than RFID tags can be generalized as nodes in wireless sensor and actuator network (WSAN), or wireless sensor network (WSN), which have been a popular research area [25, 35]. Such nodes, also commonly termed as IoT devices, are embedded systems with constrained but considerable computational power compared to the passive RFID tags. They can be fixed or portable devices, including sensor nodes installed on rooftop of buildings and wearable fitness monitors, that equip the standalone operating capability, Internet connectivity, as well as input and output peripherals for sensing and interacting with the environment and humans [39]. These constrained nodes that enable autonomous sensing and control are the elementary components in IoT applications, and hence they will be the core of the discussion in this chapter.

Cloud analytics and data visualization make use of cloud computing for processing the tremendous amount of raw data generated by the *Things* using statistical methods, machine learning and other computational intelligence algorithms, and eventually presenting them as insights for consumption by humans [39, 64]. As these tasks may involve huge amount of computational power, they are often assigned to high-performance computer clusters in data centers [17], which are often remote to the location where the IoT applications are deployed.

The enabling technologies introduced above show that heterogeneity exists in every aspect of the IoT ecosystem, due to diverse application requirements, device standards, communication protocols and other factors. As a consequence, interoperability should be the first principle in designing and maintaining IoT applications. This is especially essential in mission-critical applications, such as power grids and security-related systems, which compatibility with legacy devices, reliability and resiliency are of the utmost importance.

1.3 The Client-Gateway-Server Model

Despite the heterogeneous nature, the system architecture of IoT applications can be abstracted using the proposed *Client-Gateway-Server* model, which is extended from the well-known client-server computing paradigm. The classical client-server model describes the relationship of resource flow between a client as the service requester, and a server as the service provider [56]. Using the World Wide Web as an example, a web browser acts as a client to request a web page from a web server, then the server responds the page to the browser for displaying on the screen. The underlying communication between the web browser and the server is in the form of data packets routed across the IP network.

The previous discussion stated that IoT can be seen as a progressive paradigm shift from the established human-driven Internet. Thus, in the *Client-Gateway-Server* model illustrated in Fig. 3, IoT devices and the cloud servers in data centers act as clients and servers, respectively, similar to the above World Wide Web example. For instance, in a hypothetical "smart thermostat" application, the thermostat as an IoT device, would periodically measure the room temperature and report it in a request over the Internet to the cloud, where intelligent processes take place to optimize the temperature setpoint for energy-saving and thermal comfort purposes. The new setpoint calculated on the server would be passed back to the thermostat as a response,

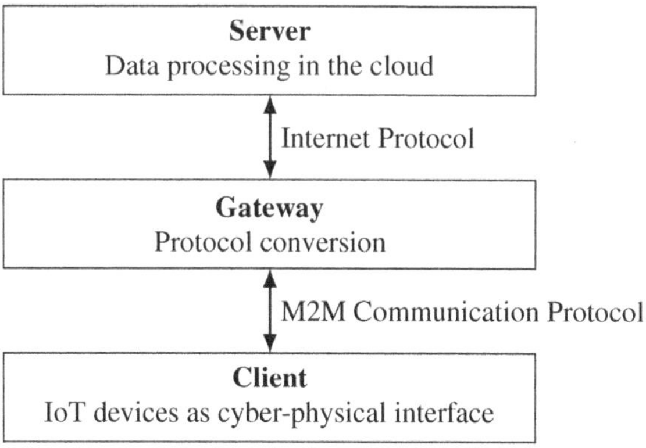

Fig. 3 The *client-gateway-server* model for the system architecture of IoT applications

and the thermostat would control the heating, ventilation and air-conditioning system accordingly.

The extra *Gateway* layer in between the *Client* and *Server* layers differentiates this newly proposed model from the classical one. The significance of this additional layer shall be emphasized due to the heterogeneous nature of connectivity in IoT devices. Unlike conventional Internet-connected devices like personal computers and smartphones, IoT devices (e.g. the "smart thermostat" in the above example) are much more resource-constrained, in terms of computational power and energy budget (i.e. battery life). As a result, many of these devices cannot afford to have a standard IP networking stack equipped. Instead, IoT devices may employ other M2M communication protocols of proprietary or open standards, such as Bluetooth, ZigBee [53] and Modbus [103], to enable data exchange with Internet servers through the use of gateway devices. Gateway devices in this context refer to wireless base stations or equivalent systems, which are responsible for performing protocol conversion to translate the packets in specific M2M protocols to standard data packets compatible with IP networks, such that the data can be routed upstream to the server, if not bi-directionally, via the regular Internet infrastructure. It should be noted that an IoT device may simultaneously acts as a client and a gateway for neighboring or companion devices. This usually happens with the adoption of mesh networks [57], and in case of connecting wearable devices to smartphones via Bluetooth [36].

1.4 Key Challenges of Internet of Things

The envisioned future of IoT presents a series of challenges yet to be solved, in both technical and social aspects. The key challenges can be prioritized into two classes. The top-priority challenges concern about the technological feasibility of IoT, which can be further divided into subproblems in the *Internet* and *Things* perspectives. The challenges regarding the *Internet* include network capacity, quality of service (QoS) considerations, M2M and information exchange protocols at different levels [39], and most of these areas are under active research. Compared to the network-related issues, the challenges in energizing the *Things* have been considered to be the bottleneck to realize IoT [21, 47], because the rate of improvement in battery and other energy storage technologies has been falling much more behind the continuous increase in the power requirements in various applications of growing complexity. This mismatch leads to the main theme of this chapter on examining possible solutions in powering IoT devices. The key challenges of relatively lower priority include the reliability and social acceptance of IoT, which should focus on addressing the security and potential privacy issues associated with this emerging Internet paradigm [49]. Other than these issues, Stankovic presented a comprehensive survey of IoT research problems in [92].

2 Energy Consumption of Internet of Things

Computational power is built upon electric power. Through referencing to the Client-Gateway-Server model introduced in Sect. 1.3, this section firstly examines the situation of energy consumption in IoT layer by layer, and then discusses the implied challenges in realizing IoT and some possible solutions.

Cloud servers reside on the top of the three-layer model. These servers are often virtual machines hosted on clusters of high-performance computers, located at data centers with high bandwidth Internet links distributed around the world [17]. The data centers are undoubtedly connected to the power grid with uninterruptible power supply to support the high energy demand of the clusters, and to ensure service reliability and resiliency [38]. More recently, the industry works towards committing to a complete adoption of renewable energy to achieve their sustainability goals, through purchasing the "green" electricity from the grid and generating on their own near the data centers [15, 106].

Gateways, in particular for the standalone base stations that serve for the conversion between IP and M2M protocols, are usually required to be fixed in close proximity to other upstream network equipment, such as routers and switches. Therefore, gateways can be powered in the same way as their neighboring equipment, possibly through the power grid or other standalone renewable energy systems. Hence, supplying electricity to standalone gateways is not a big concern in most scenarios.

Among the two types of *Things*, passive RFID tags are "self-powered" when their antennae are energized by RFID readers. In contrast to the fixed sever and gateway assets, power supply for the other category of *Things*, autonomous IoT devices with embedded computational power, is the toughest part among the three layers. Given that they may be deployed everywhere, neither using power cords nor replacing batteries regularly is a practical option for billions of such devices. Thus, the discussion in the rest of this chapter shall concentrate on this class of devices. Before introducing the possible power supply solutions, it is necessary to identify the energy consumption characteristics of IoT devices.

2.1 *Internet of Things Device Architecture*

As defined in Sect. 1.2, IoT devices refer to the broad category of embedded systems in the IoT ecosystem that have direct or indirect Internet connectivity, and the capability to interact with the physical world through on-board sensors and actuators. Although heterogeneity exists in such devices due to the diversity in the nature of applications, protocols, hardware and software design, the general architecture for IoT devices can be illustrated by Fig. 4 from the systems perspective.

Any IoT device, similar to a WSN node [25], can be considered as an integration of four subsystems, namely the processor, network interface, input and output (I/O) peripherals, as well as power supply. These four subsystems are of a system-level

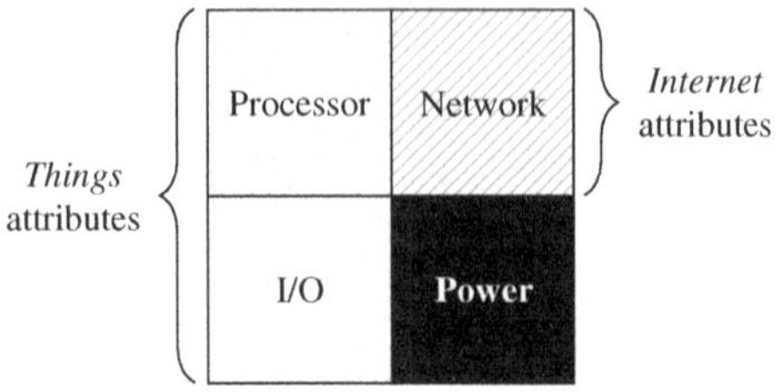

Fig. 4 General IoT device architecture

description, and they may not be four actual hardware parts in a practical device. It is not uncommon for manufacturers to integrate two or more of these subsystems into a single chip, which is known as a system on a chip (SoC).

In this architecture, the processor and I/O peripherals give the *Things* attributes to an IoT device. The processor often refers to a microcontroller unit (MCU) composed of a microprocessor aggregated with memory, timer, digital and analog I/O ports, etc. It serves as the core of an embedded system that runs software for resource management, scheduling and task execution, including the control of the I/O peripherals and the network interfaces. The I/O peripherals act as the cyber-physical interface that interact with the environment and humans, via various types of sensors and actuators. Examples include temperature sensors, accelerometers, motors, touch screens, etc. These peripherals have wired interface with either the processor's analog I/O ports, or digital ones if the peripherals have built-in analog-to-digital (mainly for sensors) and digital-to-analog (mainly for actuators) converters. The *Internet* attributes of an IoT device are provided by its network interface, which is responsible for communicating the processor to a gateway as described in Sect. 1.3, through wired or wireless connection. Since all these three subsystems require electric power to operate, the power supply subsystem is responsible for delivering power to maintain the device's operation, and for managing the energy source (e.g. mains power, battery, capacitor, etc.) to ensure safety and guarantee the designed lifetime.

2.2 Device Classification

The *Terminology for Constrained-Node Networks* (RFC 7228) published by the Internet Engineering Task Force (IETF) in 2014 [14] introduces classification schemes of IoT devices based on computational capabilities and energy limitations. However, the scheme for energy limitation classification simply separates devices into four classes by their power supply options, as tabulated in Table 1, rather than the energy consumption constraints during operations.

Jayakumar et al. also proposed another energy-related taxonomy for IoT devices in [47]. According to their work, IoT devices can be divided into five types with regard to their power and longevity requirements, as shown in Table 2. Although this longevity-based classification differentiates the ranges of device lifetime regarding

Table 1 Classes of energy limitation for constrained devices in RFC 7228 [14]

Class	Type of energy limitation	Example power source
E0	Event energy-limited	Event-based harvesting
E1	Period energy-limited	Battery that is periodically recharged or replaced
E2	Lifetime energy-limited	Non-replaceable primary battery
E9	No direct quantitative limitations to available energy	Mains-powered

Table 2 IoT device classification proposed by Jayakumar et al. [47]

	Description	Examples	Energy source	Lifetime
Type I	Wearable devices	Smartwatches Fitness monitors	Rechargeable battery	Several days
Type II	Set-and-forget devices	Home security sensors Water leak sensors	Battery	5–10 years
Type III	Semi-permanent devices	Structural monitors Parking space sensors	Battery	>10 years
Type IV	Battery-less devices	RFID tags Smart cards	Passively powered	–
Type V	Powered devices	Home appliances (e.g. refrigerators)	Power outlet	–

its application requirement, neither this scheme nor the RFC 7228 helps analyze the energy consumption pattern of an IoT device.

Therefore, a basic analytical framework for characterizing the energy consumption of IoT devices is formulated based on the *Internet* and *Things* attributes in the aforementioned device architecture. Since energy consumption of an electrical load is the integral of instantaneous power consumption over a given time interval, the IoT device energy consumption should be evaluated from both the time-domain and power-domain. The time-domain can be described by the operating mode of a device, while the power-domain can be considered in terms of the network interface power usage, which is typically much more significant than that of the processor and I/O peripherals [47]. It is anticipated that this analytical framework can aid the design and implementation of effective power supply solutions for IoT devices.

2.3 *Device Characterization by Operating Mode*

It is assumed that any application running on an IoT device can be modeled as the operating cycle illustrated in Fig. 5. When the application starts in each cycle, the processor wakes up from a low-power "sleep" mode to the "active" mode. Sensors then operate to take measurements (for devices with sensors). Next, the device connects to the gateway for transmitting the measurement data to the server as a request, and checks if there are any commands for the device to execute from the server response. After that, the actuators operate according to the received commands (for devices with actuators). At the end of the task sequence, the device turns to the power-saving "sleep" mode as it becomes idle. This model is purposed to outline the distinctive processes that lead to significant energy impact in the IoT device applications only. In practice, the processes may be executed concurrently, in a slightly different order, or on top of an operating system [43], instead of running sequentially as illustrated in Fig. 5. Further implementation details in hardware and software aspects shall be discussed in Sect. 4.

According to this operating cycle model, the operating modes for IoT devices can be classified into four patterns: *Time-triggered* and *event-triggered* modes form the basis [51], and the *always-on* and *event-blocking* modes are then derived. In order to focus on describing the general operating mode patterns, it is also assumed that the power level only toggles between the high-power "active" level, P_{active}, and the low-power "sleep" level, P_{sleep}, in this time-domain discussion. Moreover, *duty cycle* is an important concept of expressing the behavior of toggling between the two power levels. As shown in Eq. 1, the duty cycle D refers to the ratio of active duration, t_{active}, to the total cycle duration that is the sum of t_{active} and sleep duration, t_{sleep}. Assuming that t_{active} is constant, the higher the duty cycle (i.e. shorter t_{sleep}), the more frequent the device wakes up to execute the task sequence.

$$D = \frac{t_{active}}{t_{active} + t_{sleep}} \tag{1}$$

The time-triggered mode refers to periodic wake up of the device as shown in Fig. 6. Devices operating in this mode execute an operating cycle at fixed or varying duty cycles. Since the duty cycles are scheduled by the processor in either case, the energy consumed by the device can be approximated by the processor with relative ease as it can keep track of the scheduling [32]. This operating mode is

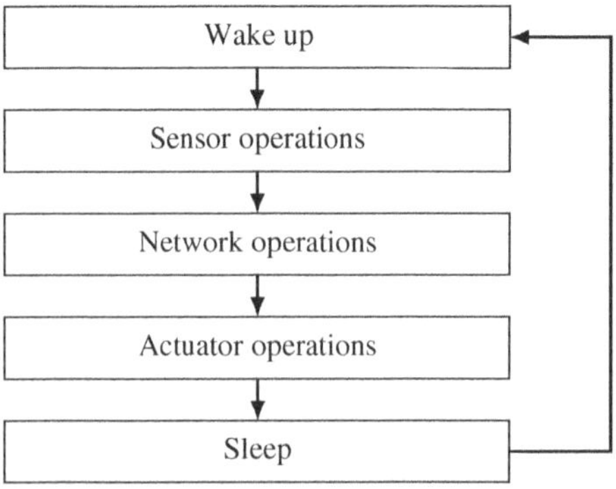

Fig. 5 Operating cycle model for IoT devices

common for autonomous devices in which human intervention is not involved, such as environmental sensor nodes that perform measurements and transmit measured data to a server regularly [2].

The event-triggered mode describes a device that is required to wake up to respond to an external event, such as the detection of a change in the environment by the device's sensor, as illustrated in Fig. 7. Since such external events occur sporadically with randomness, it may not be possible for the processor to predict the wake up in advance, and hence energy consumption estimation for this operating mode is infeasible in most cases. This type of operating mode can be found in autonomous devices including motion-activated security alarms and structural health monitoring sensor nodes activated during an earthquake [22], as well as in human-operated devices that activate upon a button push [16].

The always-on mode can be considered as a special case of the time-triggered mode, in which the device keeps repeating the task sequence and never goes to sleep (i.e. a duty cycle fixed at 1), as shown in Fig. 8. This continuous mode facilitates autonomous devices to achieve real-time streaming of sensor data to the server, as well as human-operated devices which stays active to await user input, and to output updated information to the user. As the device never sleeps, the power consumption is considered to be constant in general.

The event-blocking mode depicted in Fig. 9 postulates a special case between the event-triggered and the always-on modes. Usually found in human-operated devices like wearable devices, they are activated in response to an external event as in the event-triggered mode. The major difference between autonomous and human-operated devices is that the latter needs to wait for human input in a significantly longer and variable duration (seconds or longer), compared to the quick sensor polling in the former (often shorter than one second). As a consequence, this operating mode is characterized by increased and unpredictable energy consumption due to the extended and dynamic operating duration per cycle.

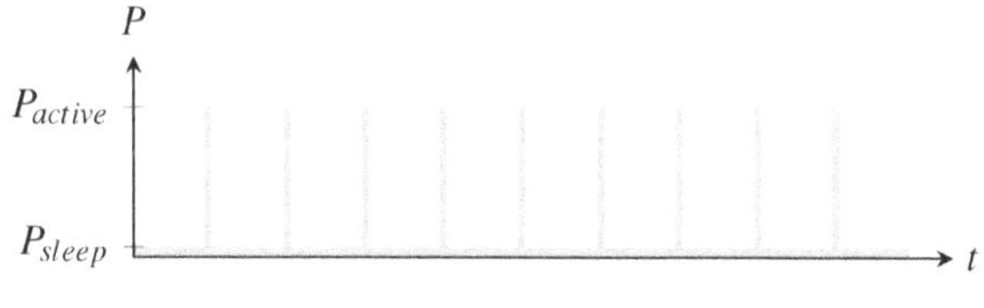

Fig. 6 Time-triggered device operating mode

Fig. 7 Event-triggered device operating mode

Fig. 8 Always-on device operating mode

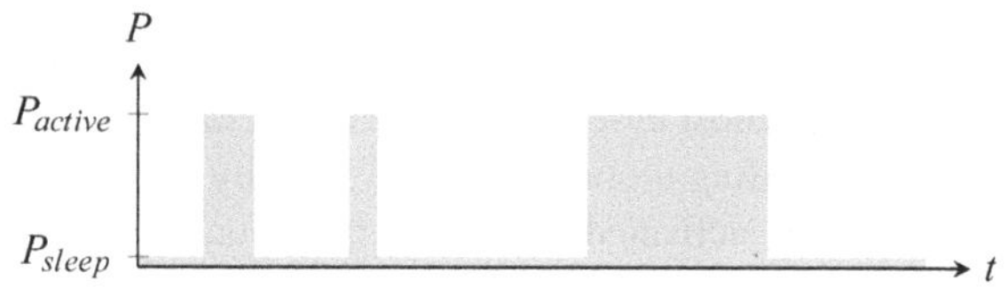

Fig. 9 Event-blocking device operating mode

2.4 *Device Characterization by Connectivity*

Based on the device architecture and the above operating cycle model, the three subsystems, including processor, I/O peripherals and network interface need to be powered up for operation as soon as the device wakes up. Among these subsystems, the network interface usually dominates the device's energy consumption [47]. Thus, the power-domain characterization of the IoT device energy consumption will focus on the communication technologies they use.

Both wired and wireless communication technologies are possible solutions for IoT devices to realize Internet connectivity, although wireless links are often preferred to enable untethered operations in remote locations. It is worthwhile to note that contrary to the human-driven Internet usage that desire low latency, high data rate networks for web browsing and video streaming applications, IoT device network traffic is typically characterized as very small volume in each transmission, delay-tolerant (except for some video surveillance and real-time sensing applications), and lower data rate is also acceptable for a majority of applications, in favor of energy-efficient networks to prolong battery life [65]. No matter using wired or wireless communications, the main role of the network interface subsystem in the device architecture is to handle network connection, with the ultimate aim to transfer data bits between the device and the gateway through a physical medium, such as copper wire, optical fibers, air, etc. This low-level bitwise communication is achieved by a transceiver, which is probably the most power-consuming hardware in an IoT device as it is required to radiate signals into air, or to drive signals in long cables.

2.4.1 Wireless Connectivity

Wireless networks can be classified based on their range and coverage, as well as their network topologies. For range and coverage, a network can be generally described as either a Wireless Personal Area Network (WPAN), Wireless Local Area Network

(WLAN) or Wireless Wide Area Network (WWAN) [65]. WPANs typically refer to short-range networks that cover within 10 meters, with Bluetooth as the most well-known technology. WLANs such as Wi-Fi, provide coverage of about 100 meters for use in an office or apartment. WWANs commonly include cellular networks that offer city-wide coverage.

Apart from the transmitting and receiving power, network topology also affects the device energy consumption. Wireless networks can typically be considered as either a star or a mesh. In the star topology, each device in the network, also known as a node, communicates with the gateway, also called the sink, via a point-to-point connection. Although this is a straightforward approach with predictably low latency, as well as the ease in design and implementation, the gateway can be a single point of failure in the system. Also, when considering the energy consumption aspect, it may incur higher peak power requirement of the radio transceiver than that of the mesh topology, provided that they have the same gateway and number of devices in a network. A mesh network is typically formed by multiple devices and one or more gateways. In contrast to the star topology, some mesh devices may serve as an intermediate gateway, known as a router, to relay the network traffic between the gateway and its neighboring devices, and this is called a multi-hop network [57]. Despite its complexity and potentially increased latency, this more flexible network configuration can reduce a transceiver's peak power of a device, since a node can communicate with another nearby node, instead of initiating radio transmission with the gateway located farther away. However, the multi-hop operations have another implication on the energy consumption profile. As a device may also act as a router, it needs to handle the additional, and probably unforeseeable, network traffic from its neighboring nodes. The overhead incurred may have considerable energy impact, depending on the technology and network configuration, and such nodes may not be clearly characterized by any of the aforementioned operating modes.

Some prevalent wireless M2M communication technologies in the IoT era, including the existing and forthcoming ones, will be briefly introduced in the following, with a focus on their ranges, data rates and network topology. Table 3 surveys various commercially available wireless M2M transceivers and provides a general idea on their relative power usage, by comparing their transmitting (Tx) and receiving (Rx) current consumption levels.

Bluetooth Low Energy (Bluetooth LE or BLE) is an example of low-power WPAN that operates in the 2.4 GHz unlicensed industrial, scientific and medical (ISM) band with a maximum data rate of 1,000 kbps, and a typical range of tens of meters [36]. BLE adopts a master-slave type of star topology, which enables IoT devices such as smartwatches and other kinds of wearable devices, to connect to a smartphone that serves as a gateway to the Internet via cellular network or Wi-Fi.

ZigBee is a low-rate WPAN (LR-WPAN) technology that is built on top of the IEEE 802.15.4 Medium Access Control (MAC) layer protocol, with a typical range of about 10–100 m [53]. This allows low-power IoT devices, typically sensor and actuator nodes, to operate under star or mesh topology with a maximum data rate of 250 kbps in the unlicensed band, using the 868 MHz, 915 MHz or 2.4 GHz spectrum [57]. Apart from ZigBee, there are also other IEEE 802.15.4-based protocols under

Table 3 Examples of commercially available wireless M2M transceivers for IoT devices

Technology	Transceiver[a]	Current consumption (mA)	
		Tx	Rx
Bluetooth LE	Atmel SAM B11[b] [7]	3 (0 dBm)	4
	Nordic Semiconductor nRF51822[b] [72]	8 (0 dBm)	10
	Texas Instruments CC2640[b] [98]	6 (0 dBm)	6
ZigBee	Atmel SAM R21 series[b] [8]	14	12
	Silicon Labs EM35x series[b] [91]	31 (+3 dBm)	27
	Texas Instruments CC2630[b] [97]	6 (0 dBm)	6
Wi-Fi	Espressif Systems ESP32[b] [31]	120 (0 dBm)	90
	Texas Instruments CC3200[b] [96]	229	59
Cellular	u-blox SARA-U2 series[c] [104]	215 (GPRS)	–
		580 (HSDPA)	
	u-blox TOBY-L100[c] [105]	460 (HSPA)	
		800 (LTE)	–
LoRa	Semtech SX1272[c] [87]	28 (+13 dBm)	10
SigFox	Atmel ATA8520D[c] [6]	33 (+14.5 dBm)	10
LEO satellites	Iridium 9603[c] [46]	145	39
		1300 (Peak)	156 (Peak)

[a] Sorted in alphabetical order
[b] SoC embedded with MCU and transceiver
[c] Standalone transceiver

active development and standardization, including IPv6 over Low-power WPAN (6LoWPAN) and Thread [20, 65].

Wi-Fi is the most commonly adopted WLAN technology based on the IEEE 802.11 standards. Operating in the 2.4 and 5 GHz ISM bands, Wi-Fi is designed to provide high speed (10 Mbps up) wireless links for devices to access to the Internet in the range of about 100 m, via access points (APs) as gateways [53] using a star topology. Despite higher power requirements of the transceivers as shown in Table 3, deploying IoT devices with Wi-Fi has a distinct advantage that these devices can utilize the well-established Wi-Fi APs in buildings and cities, hence reducing additional costs on new gateway infrastructure [101].

Conventional cellular networks using Global System for Mobile Communications (GSM), General Packet Radio Service (GPRS), Long Term Evolution (LTE) technologies provide mobile access to the Internet as WWANs using licensed spectrum. Since cellular networks are designed to optimize for high-quality voice and high-throughput data transmission with considerations for hand-off between cell towers, they require high power transceivers, and thus not a suitable choice for IoT devices in general. In view of these disadvantages, the Third-Generation Partnership Project (3GPP) completed standardization work of three tracks of cellular low-power WAN (LPWAN) in 2016: Extended Coverage GSM (EC-GSM), LTE enhancements for Machine-Type Communications (LTE-M) and Narrowband IoT (NB-IoT) [30]. These three emerging technologies are designed to cater for IoT applications with low-cost and low-power requirements, and they can fully leverage the existing GSM and LTE spectrum and base station infrastructure for rapid network deployment (only software upgrade is required) [37]. Nokia suggested that these cellular IoT technologies can offer a long communication range up to 35 (EC-GSM) or 100 (LTE-M and NB-IoT) kilometers, with the maximum data rate of 140 kbps (EC-GSM), 170 kbps (NB-IoT) or 1 Mbps (LTE-M) [71]. Although there is no commercial transceiver available to date, it has been claimed that the transceiver modules based on the above technologies will cost as low as USD 5 per unit, and they can last for 10 years under battery operations [30, 71, 113].

LoRa and SigFox are emerging proprietary LPWAN technologies using the unlicensed spectrum in the sub-GHz band [58]. Similar to the cellular IoT technologies, they offer low-power and low-cost transceivers, enabling IoT devices to achieve long-range communication with gateways kilometers apart with a star topology. This reduces the need of a large number of gateways to drive down the infrastructure costs [82].

Satellite M2M communications is an interesting class of connectivity that provides global coverage for IoT devices in remote locations, out of the reach of any of aforementioned M2M communication technologies [42]. Traditionally, accessing satellite Internet on the ground and on aircrafts requires a dish or phased array antenna pointing at communication satellites in the Geostationary Earth Orbit (GEO) at about 36,000 km above the equator, which implies prohibitive power and size requirements for off-the-grid, battery-powered IoT devices [27]. Instead of accessing Internet directly, current satellite M2M applications rely on satellite constellations in the Low Earth Orbit (LEO), which compose of a network of satellites orbiting at an altitude of 2,000 km or below in different orbital planes, to provide relatively low latency (compared to GEO) and low bandwidth message transmission via dedicated ground stations [62]. New LEO Internet constellations consist of hundreds to thousands of satellites are also in active development, and these massive constellations aim to provide high bandwidth and true global coverage [42, 69]. It is foreseeable that future autonomous IoT devices may be able to adopt relatively low-cost, low-power and small-sized transceivers to achieve Internet access via such LEO constellations.

2.4.2 Wired Connectivity

Although wireless communication is often preferred, as the ubiquitous nature of such devices renders technical or economical infeasibility of expanding the wired infrastructure coverage, the importance of wired communication in the IoT paradigm should not be undermined. In some enterprise, mission-critical applications, wired networks offer the advantages of enhanced reliability, security, higher data rate and immunity to radio interference. Examples of wired communication technologies include Ethernet, serial communication (e.g. EIA-232 and EIA-485) [63], power-line communication [41, 61] and Modbus [26]. Except for Ethernet, the other technologies require additional gateways to allow IoT devices to access to the Internet.

2.5 Power Supply Options

The above analytical framework suggests that significant variety exists in the energy consumption of IoT devices, according to their operating modes and connectivity. Therefore, power supply options have to be considered thoroughly when designing an IoT device, such that its power requirements can be satisfied. In general, the power supply subsystem of an IoT device is required to deliver power to meet the device's demand from an external source (e.g. mains electricity or standalone generators), an internal energy storage (e.g. battery or capacitor), or a combined approach that uses an external source for recharging the storage. As processors and numerous transistor-based parts are required to operate at a stable voltage, commonly being 5 V, 3.3 V or 1.8 V DC, it is needed to regulate the input source before distributing power to these components. Moreover, if the device adopts an internal energy storage, its charging and discharging has to be managed to ensure operational safety and performance. These functions are achieved in a power management unit as illustrated in Fig. 10, with the power flow directions between the supply and loads (processor, I/O peripherals and network interface) indicated. Four classes of power supply approaches will be examined, including tethered power transfer, energy storage, wireless power transfer and energy harvesting. It has to be emphasized that although the four classes of power supply approaches are introduced independently, they are not mutually exclusive. Instead, practical IoT devices often involve hybrid use of power supply options within the same class or in different classes. For instance, energy storage is routinely paired with power transfer or energy harvesting techniques to extend device lifetime.

2.5.1 Tethered Power Transfer

Tethered power transfer is a collective term for describing electric power transfer methods that involve the use of a cord connecting an external power source, for example, a mains power socket, to the device. If mains power from the grid (typically

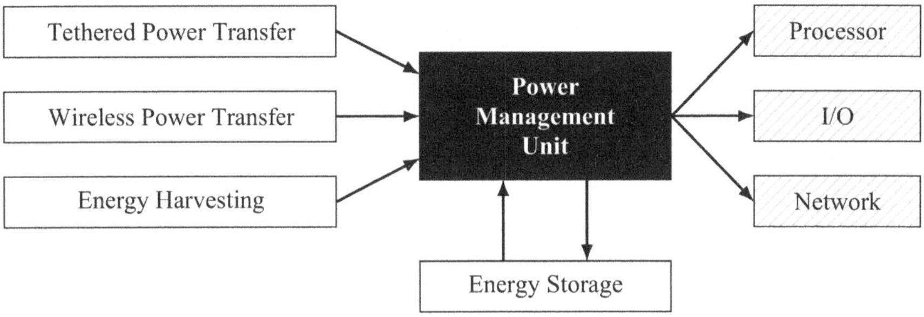

Fig. 10 Power supply flow in IoT devices

ranged from 100 to 240 V AC, depending on countries) is used on an IoT device, a bulky AC-DC converter is essential to rectify the AC power to DC, and to step down the voltage to the required level, as mentioned earlier. Apart from consuming the grid power directly, it is also possible for devices to adopt an alternative approach that utilizes DC power distribution. This can be achieved with as simple as using a pair of conducting wires extended from the rectified DC output of a central transformer, as well as more complex DC "nanogrid" standards, such as Power over Ethernet and USB Power Delivery [73]. Although tethered power transfer may not be possible for many IoT devices, for the same reasons as wired communication, tethered power transfer remains as a reliable solution to cope with high power demand, such as real-time video-streaming surveillance cameras, and recharging the energy storage of portable and wearable devices.

2.5.2 Energy Storage

Embedding an energy storage is the most popular approach in the power supply of IoT devices, given that devices are often located out of the reach of power cables. An energy storage can be categorized as either expendable or rechargeable. Expendable storage mainly refers to primary batteries that are non-rechargeable, and are mainly used in disposable and very low-power IoT devices for limited lifetime. Rechargeable storage consists of two major types, including secondary batteries, also known as rechargeable batteries, as well as capacitors. Since they are rechargeable, they can be used with power transfer or energy harvesting techniques to extend the operating lifetime. Commonly used types of energy storage are tabulated in Table 4 with their important parameters, including single-cell nominal operating voltage, typical recharge life cycles and volumetric energy density (physical size is a more critical factor than weight in most terrestrial applications). Among the different types of storage listed, lithium ion battery is the dominating rechargeable energy storage option in portable and wearable consumer electronics, electric vehicles and a wide spectrum of industrial applications, due to its high cell voltage and energy density properties [112]. More recently, lithium iron phosphate batteries and supercapactiors

Table 4 Characteristics of expendable and rechargeable energy storage [45, 48, 77, 83]

Energy storage	Cycle life	Nominal voltage	Energy density
		V	Wh/L
Primary batteries			
Alkaline	–	1.5	461
Lithium metal	–	3.0	546
Silver oxide	–	1.55	530
Secondary batteries			
Nickel-metal hydride (NiMH)	500–1,000	1.2	430
Lithium ion (Li-ion)	>1,000	3.7	570
Lithium iron phosphate ($LiFePO_4$) [45, 48]	>2,000	3.3	210
Supercapacitors [48, 77]	>100,000	2.7	4.4

have received attention and been considered as a new generation of energy storage for embedded systems and IoT devices [59, 80] despite their lower energy densities, since they offer better cycle life and electrical characteristics compared to traditional battery chemistries.

Lithium iron phosphate ($LiFePO_4$) batteries are also a type of lithium ion (Li-ion) batteries with different cathode chemistry from the commonly referred Li-ion batteries which mainly use lithium cobalt oxide ($LiCoO_2$) or lithium manganese oxide ($LiMn_2O_4$) [112]. It is reported in [59, 112] that $LiFePO_4$ is a safer, more environmentally benign option with significantly longer cycle life compared to Li-ion batteries. The enhanced safety attributes to the thermal stability of the $LiFePO_4$ cathode material, compared to $LiCoO_2$ and $LiMn_2O_4$ [119]. In addition to its safety and longevity merits, $LiFePO_4$ cells also have a favorable nominal voltage at around 3.3 V and a flat discharge curve, compared to the wide discharge voltage range from 4.2 to 3.7 V in Li-ion batteries [45]. This means that there exists a possibility for $LiFePO_4$ batteries to power 3.3 V devices directly to eliminate the overhead of an extra voltage regulator.

Supercapacitors, also known as ultracapacitors and electric double-layer capacitors (EDLC), inherit the electrical properties of capacitors like high power density and theoretically unlimited recharge cycles, but with significantly higher capacitance in the order of hundreds of millifarads to dozens of farads [77]. Compared to secondary batteries, the energy density of supercapacitors is substantially lower, and the self-discharging rate (leakage current) is much higher. Hence, supercapacitors are typically used in ultra-low power IoT devices, or in a hybrid manner with batteries to cope with high pulse load of radio transceivers [48].

2.5.3 Wireless Power Transfer

Wireless power transfer (WPT) refers to the transfer of electric power from a charging node, which serves as the power source, to a power receiving node, which is an IoT device in this context, without any physical contact between the nodes. According to Xie et al. [116], there are three categories of WPT technologies, including inductive coupling, electromagnetic (EM) radiation and magnetic resonant coupling.

Inductive coupling is a mature centimeter-range WPT technology that works by magnetic field induction. This technology can be found in wireless toothbrush and the *Qi* wireless charging pads for portable electronic devices [116]. Due to its extremely short range and requirement for accurate alignment between the charging and power receiving nodes, inductive coupling is considered to be only viable for eliminating the need for power cords when recharging the batteries of portable- and wearable-type IoT devices, rather than other autonomous devices.

EM radiation WPT relies on an antenna to emit and receive power in the form of EM waves on the charging and power receiving nodes, respectively. Two types of EM radiation WPT, namely the omni-directional and the unidirectional types, can be classified according to the direction of receiving the radiated energy. The omni-directional type receiving node only requires a tiny antenna to receive incoming waves from any direction. The receiver antenna can convert low-power propagating EM waves in the ISM frequency bands, in the range of centimeters to several meters, to energize devices with ultra-low power requirement. Passive RFID tags discussed in Sect. 1.2 are common examples utilizing EM waves emitted by the RFID reader to transfer power to the tags, in order to facilitate their data exchange processes. Low-power, autonomous IoT devices, such as sensor nodes can also potentially be powered by omni-directional radiation. On the other hand, the unidirectional type requires line-of-sight transmission. Unidirectional radiation WPT is usually designed for high-power, kilometer-range applications with the use of large-scale microwave or laser beam receiver. Thus, the unidirectional type cannot be considered as a possible power supply option for IoT devices.

Magnetic resonant coupling is a WPT technology developed more recently. First introduced in 2007 by Kurs et al. [52], this relatively new technology relies on non-radiative magnetic resonance induction to improve efficiency and power level in the meter-range drastically, compared to the two aforementioned omni-directional technologies. Once this technology becomes mature in the future, it can potentially benefit to more stable and effective power supply for various types of IoT devices, such as realizing battery-less autonomous sensor and actuator networks, and simultaneous fast recharging of multiple devices' batteries.

2.5.4 Energy Harvesting

Energy harvesting represents the extraction of energy from the ambient environment and its conversion to electricity. Since many IoT applications, for example, environmental monitoring and building automation, require autonomous devices with limited energy storage size to be deployed to locations beyond the reach of tethered or wireless power transfer infrastructure, in situ energy harvesting has been deemed a promising power supply solution to prolong the lifetime of such constrained devices, and to reduce cost and enhance safety by using small-capacity energy storage like $LiFePO_4$ batteries and supercapacitors. In the next section, the general principles and sources of energy harvesting will be examined. Design considerations for adopting energy harvesting in IoT devices will also be covered in Sect. 4.

3 Energy Harvesting Principles

Energy harvesting (EH), also known as energy scavenging, refers to the conversion of ambient energy from the natural or artificial environment into electricity. As discussed in Sect. 2, while various types of wireless M2M communication technologies have already been put in place to support billions of IoT devices, providing reliable power supply for the autonomous devices deployed in remote locations remains as a major hurdle. Given its ability to generate electricity and replenish energy storage on site, EH is considered to be a prominent solution to this bottleneck, such that the lifetime of these devices can be extended from months to years and decades, and ultimately enable their self-sustaining operations. In this section, the commonly used EH technologies, and the principles required to understand the design and operations of EH-powered systems will be introduced.

3.1 Energy Harvesting Technologies

Diverse technologies of EH transducers can be used in converting different forms of energy into electricity. In general, EH transducers for autonomous IoT devices can be classified as radiant, mechanical, thermal or magnetic type [77], according to the form of energy in which the source exists. The following provides an overview of commonly adopted transducers, and Table 5 shows their applicability in IoT devices found in the literature. Working principles and technical details of transducers can be found in [79].

Radiant energy, mainly the propagating electromagnetic waves of different wavelengths, can be harvested as electricity. Photovoltaic (PV) cells, also widely known as solar cells, can convert visible light into electricity using the photovoltaic effect. Solar and indoor light energy harvesting has been a common approach for powering autonomous sensors [19, 74, 81, 120]. Radio-frequency (RF) energy harvesting generates electricity by using an RF antenna to capture energy from radio signals [75, 108], in a way similar to EM radiation WPT introduced in Sect. 2.5.

Mechanical energy can be used to generate electricity. Motions in natural and artificial environments can be converted into electrical energy with three transduction mechanisms: Piezoelectric, electromagnetic and electrostatic [10]. Piezoelec-

Table 5 Examples of energy harvesting transducers for IoT devices

Energy	Transducer	Possible application scenario
Radiant	Photovoltaic	Solar-powered wireless sensor node [74, 81]
		Indoor light-powered wireless sensor node [19, 120]
	Radio-frequency	Wi-Fi signal-powered digital temperature meter [108]
		Radio signal-powered wireless sensor node [75]
Mechanical	Piezoelectric	Vibration-powered wireless structural health monitor [70]
	Electromagnetic	Vibration-powered wireless sensor node [99]
		Wind-powered wireless sensor node [74]
	Electrostatic	Vibration-powered battery charging [100]
Thermal	Thermoelectric	Human body heat-powered wearable sensor [55]
		Thermal gradient-powered wireless sensor node [28, 94, 111]
Magnetic	Coil	Current transformer-powered wireless sensor node [86]

tric generators use piezoelectric materials, such as lead zirconate titanate (PZT) and polyvinylidene fluoride (PVDF), to convert mechanical strain into electric power. Vibrations generated by machines and human walking can be harvested by piezoelectric generators [70, 93]. Electromagnetic transducers generate electricity using Faraday's law of electromagnetic induction. Wind turbines are examples of this type of transducers [74]. Electrostatic transduction refers to the conversion of mechanical energy into electricity as the capacitance of a variable capacitor changes when its charged plates displace under vibrations [100].

Thermal energy, which refers to a temperature difference in this context, can be harvested using a thermoelectric generator (TEG). TEGs are solid state devices that utilize the Seebeck effect to generate electricity when they are subjected to a temperature gradient [10]. Example applications include powering wearable devices [55], terrestrial [28, 111] and satellite [94] sensor networks.

Magnetic energy harvesting is possible via harnessing electricity from varying magnetic fields using coils. This can be applied to extracting electrical energy from AC power cables using current transformers. This technique has a large potential to aid power utilities to deploy sensors along the grid for condition monitoring of overhead and transmission lines in a non-invasive manner [86, 121].

3.2 Energy Source Characterization

Section 3.1 suggests that various EH technologies can be employed to capture energy from the natural and artificial environment. As summarized in [93, 109] and shown in Table 6, EH sources have very low power output in essence. Apart from its amplitude, the harvestable energy from an energy source is also constrained by its temporal availability. According to Kansal et al., an EH source can be characterized by its *controllability* and *predictability* [50].

Intuitively, controllability refers to the ability whether the application designer can control the energy yield from a source, including its occurrence and magnitude. Natural sources, such as sunlight, wind and tides, are obviously uncontrollable sources. Artificial sources like indoor lighting and RF, may be either controllable or uncontrollable. For example, if an RF transducer attempts to harvest the background EM waves emitted by cell towers in the surroundings, this source is considered as uncontrollable since the application designer has no authority to control the radio transmission of the cellphones. On the other hand, in case of the heat dissipated from the radiator of a compressor unit can be treated as controllable, since it is known that a certain amount of energy will be available as soon as the unit starts. Hence, a controllable source also implies a predictable one.

Predictability concerns about whether the behavior of a source, which mainly refers to its availability and magnitude, can be practically modeled with reasonable accuracy. Sunlight is a predictable source, since its amplitude and long-term availability at a certain location is fixed and can be calculated according to the well-understood celestial mechanics of Earth and the Sun, and its short-term fluctuations in the prediction can also be complemented with weather forecast. On the contrary, unpredictable sources involve uncertainties and random events, such as mechanical vibration due to earthquakes and strain induced by impact between objects and structural failure.

Table 6 Typical power densities of common energy harvesting sources [93, 109]

Source	Harvestable power density ($\mu W/cm^2$)
Light	
Solar	100,000–15,000
Artificial	10–100
Vibration and motion	
Human body	4
Industrial	100–200
Thermal gradient	
Human body	30
Industrial	1,000–10,000
Radio-frequency	
Cell tower	0.1

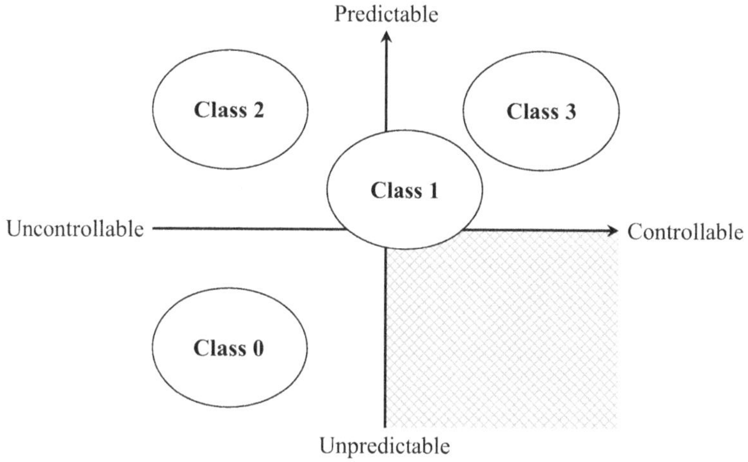

Fig. 11 Classification of energy sources according to controllability and predictability

Based on the controllability and predictability criteria, an energy source can be characterized in these two dimensions with four classes as illustrated in Fig. 11 and discussed below, with example scenarios listed in Table 7.

Class 0 energy sources are uncontrollable and unpredictable. An uncontrollable and unpredictable source means that its occurrence can neither be controlled by the application designer nor predicted by the EH system without complex models, as discussed above. This category contains natural sources, including vibration during earthquakes, and rarely occurring man-made sources subjected to randomness, such as vibration caused by the impact forces between vehicles in a traffic accident, where the patterns are difficult to formulate. This kind of sources are considered impractical, if not impossible, to be harvested to power up IoT devices operating under the more energy-intensive time-triggered, always-on and event-blocking modes. Class 0 sources are of an unstable nature without repetitive patterns. However, there exists the possibility that Class 0 sources can be used as the power supply, at the same time as an interrupt signal, for some event-triggered devices.

Class 1 refers to partially controllable sources. A partially controllable source refers to one that its availability cannot be fully controlled by the application designer. This category of sources includes RF energy incurred by spontaneous transmission associated with cellphone usage as mentioned earlier. Energy harvesting may be possible under this sort of scenarios, but the amount of energy extractable may be considerably smaller than fully controllable sources. Therefore, only ultra-low power IoT devices operating under the time-triggered and event-triggered modes may adopt this class of sources. Special power management strategies may be needed for such devices, for instance, wake up only if the energy storage level is sufficient for the device to operate for one complete cycle.

Table 7 Example scenarios of energy harvesting for IoT devices with Classes 0–3 energy sources

	Scenario	EH transducer	Operating mode
Class 0	Impact-activated structural sensors	Piezoelectric	Event-triggered
Class 1	Low-power environmental sensors	RF	Time-triggered
Class 2	Solar-powered irrigation system	Photovoltaic	Time-triggered
Class 3	Power transmission cable monitoring sensors	Current transformer	Always-on

Class 2 sources are uncontrollable but predictable. In contrast to Class 0, Class 2 sources can be forecast with simple models or based on historical patterns, and thus can be practically harvested with higher reliability on energy yield. Most of the natural, renewable energy sources, including solar, wind and tidal energy fall into this category. For example, solar and wind energy of a certain time at a location can be predicted using seasonal observation data supplemented by weather forecast. Given its higher energy yield, this type of energy sources can potentially replenish charges in devices' energy storage faster and support wider range of operating modes with higher duty cycles.

Class 3 describes fully controllable sources, which are ideal sources in the sense that their availability and output can be determined in advance of device design. Examples of energy sources include the heat and vibrational energy come from the compressors of air-conditioning systems and other machines. In this case, the yield of energy harvesting can be effectively approximated for powering various types of IoT devices of all operating modes. In addition, designs without energy storage are possible with proper modeling of the devices' energy consumption.

3.3 Energy Harvesting Architectures

There are two typical types of architectures for EH-powered systems, known as Harvest-Use and Harvest-Store-Use [93]. They provide the basis of device design that enable the utilization of the small power output from EH sources with different availability patterns.

The Harvest-Use architecture directs the electrical energy from the EH unit, which consists of the transducer and a power converter, to power the electrical load as depicted in Fig. 12. Since energy storage is absent in this architecture, it is necessary for the EH unit to supply power to the load on its own. In other words, the power delivered to the load has to be higher than the load's minimum power requirement at all time during operations. This contributes to the major disadvantage of the Harvest-Use architecture and renders it an impractical architecture for Classes 0–2 energy sources discussed in Sect. 3.2, as it is sensitive to variation in the EH unit's power output. This architecture may only be useful for certain applications adopting Class 3 energy sources.

In contrast to the Harvest-Use architecture, the Harvest-Store-Use architecture introduces an energy storage device in between the power flow from the EH unit to the load, as illustrated in Fig. 13. The energy storage, typically one or more rechargeable batteries or supercapacitors, enables continuous operations of the load in two ways. First, the storage can serve as an energy buffer to maintain stable power delivery to the load in case of a temporary drop in the EH unit's power output. Second, the storage can be an energy reservoir to allow the use of Classes 1 and 2 energy sources. For instance, the storage can be recharged with solar energy harvesting during daytime, and supply power to the load at night. Applications with Classes 0 and 3 sources may also favor this architecture as the storage, probably a capacitor, can buffer the harvested energy and stabilize the power output to the load over the operating duration.

3.4 *Energy Neutrality*

The main purpose of adopting EH in IoT devices is to utilize the tiny amount of electricity generated by the transducers to extend the service life of the devices, and ideally to achieve self-sustaining, "perpetual" operations from the energy point of view. Accomplishing self-sustainability requires the concept of energy neutrality, which means that the harvested energy always equals or exceeds the energy usage. The energy neutrality requirements and implications of the two previously introduced EH architectures, Harvest-Use and Harvest-Store-Use, will be discussed using the mathematical models proposed by Kansal et al. [50] with some simplification. The

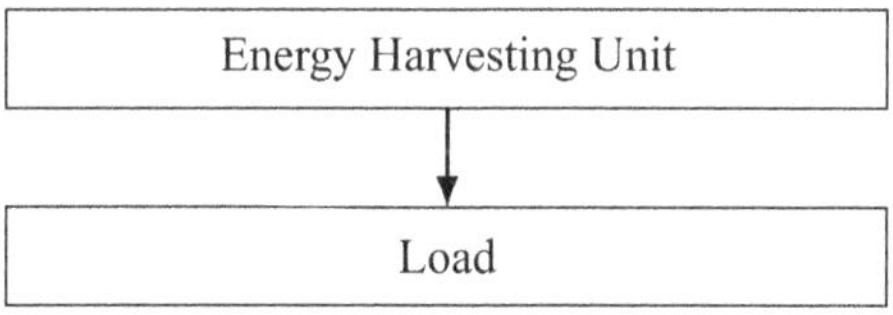

Fig. 12 Harvest-Use energy harvesting architecture

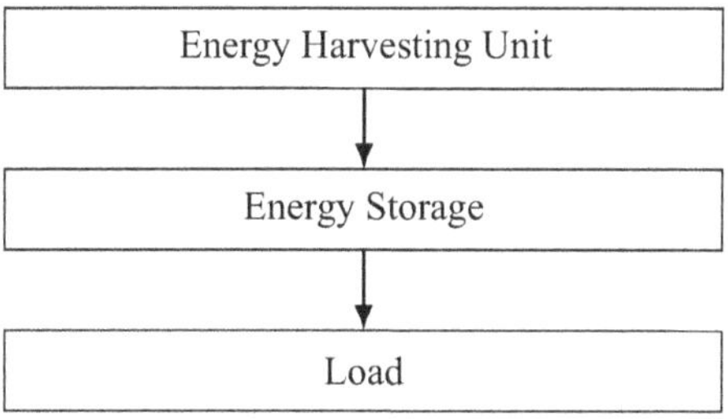

Fig. 13 Harvest-Store-Use energy harvesting architecture

power output of an energy harvesting unit and the power consumption requirement of the load at time t are notated as $P_{harvest}(t)$ and $P_{load}(t)$, respectively.

For an EH system adopting the Harvest-Use architecture without an energy storage, as mentioned in Sect. 3.3, requires that the EH power output always equals or exceeds the power input requirement to achieve energy-neutral operation, as denoted in Eq. 2.

$$P_{harvest}(t) \geq P_{load}(t)\ \forall\, t \tag{2}$$

Due to the lack of an intermediate energy storage, at the time when $P_{harvest}(t) < P_{load}(t)$, the system cannot be powered up and the harvested energy is wasted. On the other hand, if $P_{harvest}(t) > P_{load}(t)$, the excess energy cannot be used as well.

For the Harvest-Store-Use architecture, in which an energy storage is present, the criteria for energy neutrality become different. The storage, with an initial energy E_0, serves as both a buffer to deliver electric charges to the load at times of low harvester output, and as a reservoir to store excess charges. Thus, assuming an ideal energy storage with infinite capacity and no leakage, the condition for energy-neutral operation of this architecture can be modeled as Eq. 3.

$$E_0 + \int_0^T P_{harvest}(t)\mathrm{d}t \geq \int_0^T P_{load}(t)\mathrm{d}t\ \forall\ T \in [0, \infty) \tag{3}$$

This idealistic model enables an EH system to achieve energy-neutral operation even at the time that $P_{harvest}(t) < P_{load}(t)$, as long as the storage is charged and is capable of delivering power to the load.

However, a practical energy storage is not ideal, where it has a finite capacity E, an efficiency η lower than 100% and energy leakage at the rate of $P_{leak}(t)$. Under these circumstances, the modeling for energy-neutral operation becomes Eq. 4.

$$E \geq E_0 + \int_0^T \eta P_{harvest}(t) - P_{load}(t) - P_{leak}(t)\mathrm{d}t \geq 0\ \forall\ T \in [0, \infty) \tag{4}$$

This model implies that when designing an EH system using the Harvest-Store-Use architecture, it is essential to consider the constraints of the EH unit and the storage, apart from satisfying the requirements of the load demand profile, in order to guarantee energy neutrality. In addition, the introduction of the energy storage to the system opens up possibilities that an IoT device's performance can be improved at runtime, for instance, increasing the device operating duty cycle to utilize the excess harvested energy in the storage [93].

4 Designing Energy Harvesting for Internet of Things

The general architectures and principles related to the energy consumption of IoT devices and EH techniques have been studied in Sects. 2 and 3. Deriving from those

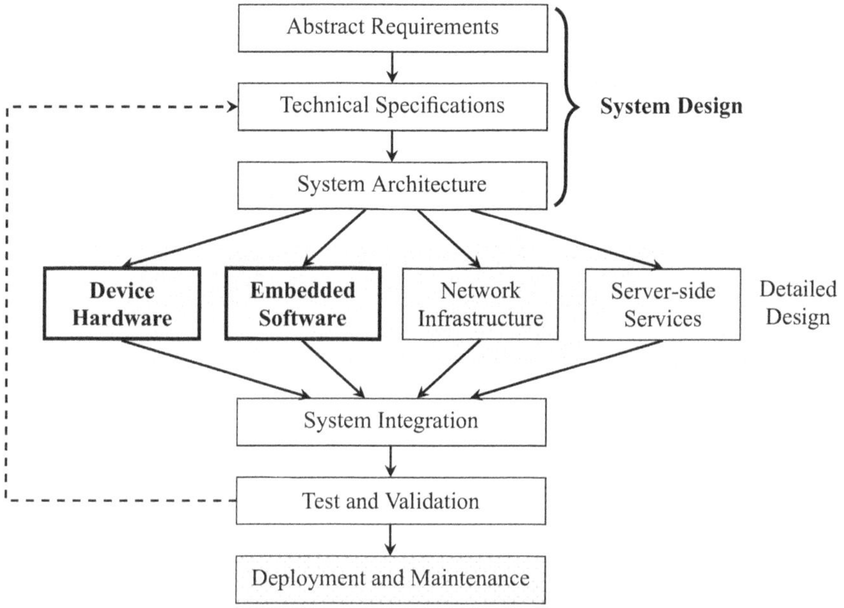

Fig. 14 Suggested development life cycle for IoT applications

concepts, some practical design considerations and guidelines for adopting EH in IoT devices will be provided in this section, with a focus on autonomous wireless sensing devices. Since such type of IoT devices resemble many design features of wireless sensor network (WSN) nodes, and embedded systems in general, literature in these fields will be examined comprehensively to aid the explanation of the design process.

Similar to software engineering, the IoT application development life cycle contains three iterative and tightly coupled phases, namely development, deployment and maintenance [13]. This engineering process requires interdisciplinary effort that involves expertise in both domain and technical knowledge [76], in order to map business requirements to application solutions that operate with IoT devices and cloud services. Figure 14 outlines a suggested design and implementation workflow for EH integration in IoT devices using a top-down approach, inspired by the established approaches for embedded system and IoT application development life cycle [11, 76, 115].

The rest of this section will focus on addressing the important issues primarily related to EH incorporation in developing an IoT device from the systems, hardware and software perspectives. The design process of an energy harvesting Wi-Fi sensor node prototyped by the authors, as shown in Fig. 15, will be used as an example for illustrating the suggested workflow.

Fig. 15 Energy harvesting Wi-Fi sensor node prototype

4.1 System Design Perspective

As suggested in Fig. 14, the system design perspective includes the initial development stages of gathering abstract application requirements, refining technical specifications and designing system architecture.

The top-level design objectives of any system are determined by some abstract requirements. In this context, requirements can be considered as either functional or non-functional [11]. Intuitively, functional requirements represent the basic functions of the system concerned, in other words, *what* the system has to do in order to solve a particular problem. Non-functional requirements refer to other factors that impose constraints on the system design and implementation, such as project budget, deployment location, environmental and legal compliance.

Upon analyzing the abstract requirements, it is necessary to translate them into more concrete technical specifications that can accurately and precisely document the verifiable attributes and behaviors of the system [11], most critically being its required inputs and outputs, physical constraints of the deployment location, allowable physical dimensions and cost. In other words, the technical specifications describe the system's external interface with the environment and the customer requirements. For applications that desire to adopt EH in IoT devices, it is also necessary to survey the intended deployment site to identify potential candidates of EH sources at this stage, by understanding the surrounding environment and measuring the variation patterns of solar irradiance, wind speed, thermal gradient, background RF signal intensity, etc.

Based on the technical specifications compiled, a system architecture, which governs *how* the system should be designed and implemented, needs to be created to define the hierarchy of interaction between smaller subsystems and components. In contrast to the technical specifications, the system architecture focuses more on the internal organization of the system. On the IoT device level, the architecture proposed in Sect. 2.1 may assist designers to analyze the required power and data flow between subsystems to achieve the desired inputs and outputs. Through establishing the system architecture, the design team may collectively decide the networking protocols, negotiate the choice of development frameworks or platforms, as well as reach consensus on the application logic and the application programming interfaces (APIs) for information exchange between services, in order to ensure that the mandatory requirements in the technical specifications can be met.

On the system architecture level, there are two areas that entail extra attention when adopting EH in IoT devices: Application logic and network protocols used. As explained in Sect. 2.3, the operating mode is crucial in determining the energy consumption of an IoT device. Thus, designers have to analyze the requirements and specifications to devise the operating mode with the minimal possible energy consumption, and the needed parameters, such as the operating duty cycle in time-triggered devices, and the exact conditions for activating event-triggered devices. Network communication also constitutes to a significant energy consumption impact, as discussed in Sect. 2.4. Hence, low-power networks should be of a high priority when making the decision about selecting protocols, provided that the physical limitations of the deployment site and budget constraints can be satisfied. Apart from the transceiver power considerations (i.e. physical and data link layer protocols), application layer protocols should also be selected to further reduce data transfer overhead and hence energy consumption. For instance, the Constrained Application Protocol (CoAP) and MQ Telemetry Transport (MQTT) are options with reduced overhead, compared to conventional Hypertext Transfer Protocol (HTTP) [3]. In addition, connection and data transmission timeout should also be configured, so as to avoid the device repeatedly retrying and draining an excessive amount of energy in the event of network failure.

The aim of the aforementioned example design shown in Fig. 15 involves demonstrating temperature and humidity monitoring in a laboratory via the Internet using a self-powered sensor node. As only hourly measurement was needed in this application, the node can be configured in the time-triggered operating mode with a low duty cycle. Since the laboratory was already equipped with Wi-Fi APs, it was planned to utilize this infrastructure instead of investing in additional networks, despite its relatively high transceiver power requirement outlined in Sect. 2.4. Harvesting solar energy using a PV panel next to the laboratory's window was considered as the only viable EH option to sustain the operations of the Wi-Fi transceiver on the node. To further reduce power requirement, the node was designed to adopt MQTT to transmit the measured data to a cloud service, where the data logging and visualization applications are hosted. The hardware architecture of the prototype is illustrated in Fig. 16.

The establishment of a system architecture facilitates the separation of concerns, where the subsequent detailed design tasks can be partitioned and assigned to different developers according to their areas of expertise [11]. It is considered that for an IoT application development cycle, the detailed design tasks can be categorized into at least four types, namely *device hardware*, *embedded software*, *network infrastructure*

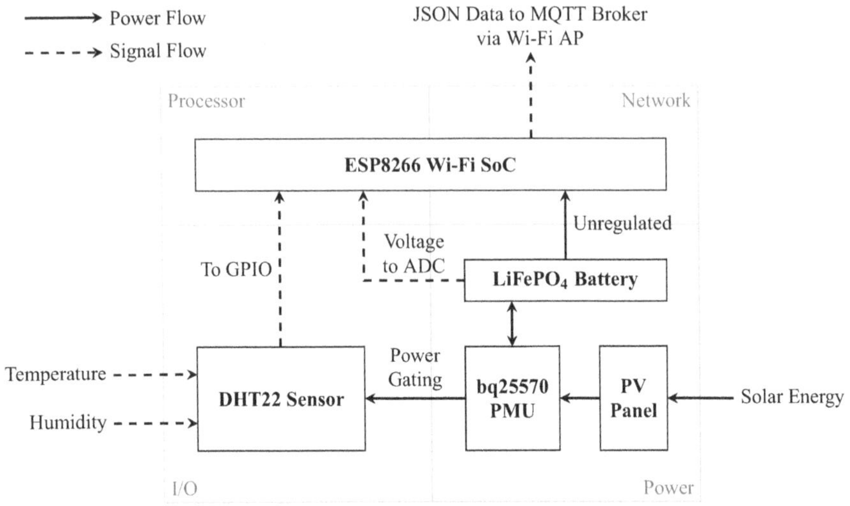

Fig. 16 Block diagram of the energy harvesting Wi-Fi sensor node prototype

(i.e. gateways and upstream IP networks) and *server-side services*, as shown in Fig. 14. With regard to the above four areas, Sects. 4.2 and 4.3 will primarily focus on addressing the detailed issues in implementing EH on IoT devices in the device hardware and embedded software aspects, while the rest of detailed design types are related to areas beyond the IoT devices and deemed out of scope of this chapter. Before deploying the application, the detailed design will eventually be integrated as a complete solution, tested and validated against the specifications to ensure that the solution satisfies the requirements, and any necessary modifications shall be made iteratively.

4.2 Device Hardware Perspective

Minimizing the power usage of an energy-harvesting IoT device is the key to achieving energy-neutral operation. Although the choice of network communication transceiver may be restricted by the application requirements, low-power processors with sleep mode options, and low-power I/O peripherals should always be preferred. After selecting the main components for the device, auditing the energy consumption pattern of the device per operating cycle is necessary, as there are actually variations in power consumption within a cycle, as opposed to the simplified two-state model introduced in Sect. 2.3. Typically, the power consumption within a cycle can be modeled with four distinct states at different times of the device operation [66]. More complex power consumption modeling methods can also be found in [43, 110].

Figure 17 illustrates the power consumption variation during an operating cycle with the four states. At t_0, the device wakes up with its processor and I/O peripherals switch from sleep to normal operating mode. This process that consumes P_{wake} typically takes only a short time, and then at t_1, the network interfaces also starts and attempts to establish a connection with the gateway, and operates in the "receiving mode" consuming P_{rx}. At the same time, the processor can also obtain required sensor data and perform any needed computations to process the data until t_2. After finishing the sensor operations and local computations, the network interface switches to the higher power "transmitting mode" consuming P_{tx} at t_2 to complete the necessary transmission. Upon successful transmission, the network interface switches back to the receiving mode and waits for the response from the server, and then the device performs the required actuator operations. When the device completes all these scheduled tasks, it turns to the sleep mode again by hibernating the processor, I/O peripherals and network interface to achieve the minimal power consumption level P_{sleep}. In reality, instead of maintaining constant power levels, those states might have slightly varying instantaneous power consumption, especially for P_{tx} and P_{rx} where short, higher-power bursts may take place occasionally during network activity.

4.2.1 Power Consumption Profile

The above paragraph introduces the variation of system-wise power consumption level, due to the heterogeneous configurations and conditions of different components in the system. This leads to the need of formulating a power consumption profile to estimate the overall energy consumption throughout an operating cycle. In practice, this can be accomplished by measuring the operational characteristics of a prototype device or through computer simulation to recognize patterns of how components change their modes during the cycle [66]. With a proper power consumption profile, the energy requirements of an IoT device, in particular the energy consumption per operating cycle, E_{cycle}, can be evaluated using Eq. 5. Different active power states (e.g. P_{wake}, P_{rx} and P_{tx} illustrated in Fig. 17) and their corresponding durations within each cycle should be taken into account in the summation of the equation.

$$E_{cycle} = E_{sleep} + E_{active} = P_{sleep}\, t_{sleep} + \sum P_{active}\, t_{active} \tag{5}$$

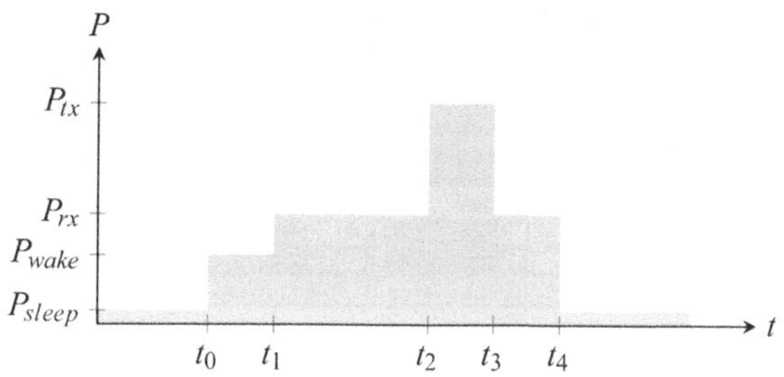

Fig. 17 Illustration of device power consumption during an operating cycle (not in scale)

The Wi-Fi sensor node prototype consists of an Espressif Systems ESP8266 SoC, which embeds a 32-bit MCU and Wi-Fi transceiver, and a DHT22 digital temperature and humidity sensor, as shown in Fig. 16. The current consumption (with constant 3.3 V power supply) during the node's operations with the required program code was inspected using an oscilloscope, as shown in Fig. 18. The power consumption profile of the node is tabulated in Table 8. It should be noted that the operating cycle model of the prototype node is not identical to the one illustrated in Fig. 17. Instead of waking up with a low power consumption, the Wi-Fi transceiver attempts connection to an AP at the beginning, and thus a power surge can be observed in Fig. 18.

4.2.2 Energy Harvesting Transducer and Storage Size Calculation

The power consumption profile establishes the basic energy requirements of the device in active and sleep phases. Next, the EH transducer and storage sizes (assuming a Harvest-Store-Use architecture) should be selected, in order to fulfill the specified device lifetime, cost and size requirements, as well as to satisfy the energy neutrality constraint introduced in Sect. 3.4 to avoid unexpected device downtime or failure. The EH source power output, and hence the amount of harvestable energy can be estimated using the methodologies that have been studied in the literature [44, 79]. The minimum required capacity of the energy storage can then be approximated by putting the device energy consumption and the amount of harvestable energy into

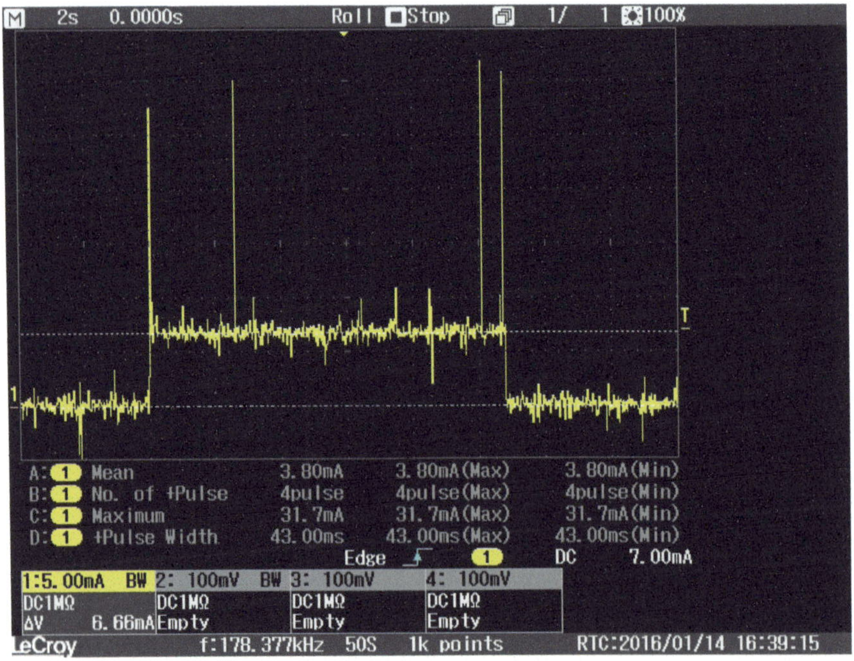

Fig. 18 Current consumption of an operating cycle of the Wi-Fi sensor node prototype

Table 8 Power consumption profile of the Wi-Fi sensor node prototype

Stage	Duration[a]	Approximate power	Energy consumption
Sleep	3591 s	190 μW	$E_{sleep} = 682$ mJ
Active	8320 ms		$E_{active} = 2.35$ J
Wake up	320 ms	860 mW	273 mJ
Sensor and network operations[b]	8000 ms	260 mW[c]	2.08 J
Full cycle	~3600 s		$E_{cycle} = 3.03$ J

[a] Average value
[b] P_{rx} and P_{tx} are similar in this case
[c] Compensated for high power bursts during transmission

balance. Sufficient design margins should be added, such that the device can survive even in the worst-case scenario, when the EH source in use becomes unavailable for an extended period [50].

For instance, according to Table 8 and Eq. 5, the prototype node consumes approximately $3.03 \times 24 = 72.72$ J daily to operate as designed to measure and transmit ambient temperature and humidity every hour. Generally speaking, in order to achieve energy neutral operations, the PV panel and energy storage have to be sized in the way that they can harvest and store 72.72 J plus the incurred energy leakage every day on average. Moreover, to ensure that the prototype node can sustain during cloudy days, which is common in the deployment location in Hong Kong, a larger energy storage is also needed to act as an energy reservoir. For this application, a 0.7-Wh $LiFePO_4$ battery is paired with a PV panel rated at 1 W, which was tested to be only able to harvest solar energy at around 6 mW on an average day. Assuming that the PV panel is capable of recharging the battery at this rate for 4 hours each day, the harvested energy ($6\,\text{mW} \times 4\,\text{h} \times 3600\,\text{s} = 86.4\,\text{J}$) can marginally achieve the energy neutrality goal. On the other hand, the 0.7-Wh battery, which is ideally equivalent to a storage of 2520 J when fully charged, can also power the prototype node for about a month ($2520\,\text{J} \div 72.72\,\text{J/day} = 34.6$ days). This example outlines a simple method for estimating the size requirements of EH transducers and energy storage, based on the power consumption profile of an IoT device.

4.2.3 Design Considerations of Power Management Unit

In general, the average EH transducer output current (and also the voltage for many transducers) is very small in nature, such that it is impossible for charging an energy storage directly with the transducer current, or delivering such unstable power to an IoT device for direct consumption. Any practical EH system would require a power management unit (PMU) to perform power conversion and energy management. In the practical design of an IoT device, PMU usually exists as a dedicated integrated circuit (IC) package, such as Texas Instruments' bq25570 [95] used in the prototype node. The PMU IC interfaces the EH transducer and energy storage for managing the power flow. It is also common for PMUs to provide energy availability indication to the processor. Figure 19 shows that the PMU possesses several roles in turning the intermittent power from an EH transducer, to stable power consumed by the load. In the following, each of these features will be briefly introduced.

EH transducers may output DC (e.g. PV and TEG) or AC (e.g. RF and piezoelectric) power. AC transducer current has to be first rectified to DC typically through a diode rectifier. The most critical role for the PMU is to use a switching-mode DC-DC boost converter to step up the millivolt-scale input voltage to a higher level, or a buck converter to step down the voltage, such that the harvested power can be converted to an appropriate DC voltage level. This is usually aided with maximum power point tracking (MPPT), which is a technique for maximizing power transfer by adjusting the power flowing from the transducer to the boost converter at an optimal point [80]. The stable output power from the boost converter is then used to charge the attached energy storage, which is a rechargeable battery, supercapacitor or a hybrid use of both, through an internal battery management system (BMS). The BMS is responsible for managing and protecting the charging and discharging of the storage, mainly through monitoring whether the storage voltage level exceeds or drops below safety thresholds. Since the nominal voltage of the energy storage may not be the same as the required voltage level of the processor, I/O peripherals and network interface, a switching-mode output voltage regulator is needed to further regulate the voltage level before distributing power to these components. Moreover, the PMU may also provide load switching capability to allow power-gating, which means that some of the loads (e.g. sensors and actuators) can be disconnected from the power supply when they are not needed to operate, through controlling by the processor [4, 47]. Since such loads contribute to some considerable energy impact in the long term, eliminating these energy overheads can potentially reduce the energy storage and even EH transducer requirements for cost saving and a simpler design.

The selection of a suitable PMU for integrating EH in an IoT device is a crucial step, as different PMUs might have different ranges of allowable input power, boost converter and BMS threshold level parameters that are optimized for certain types or models of EH transducers. There are several key reference parameters for application designers to look for and to design around during the development process.

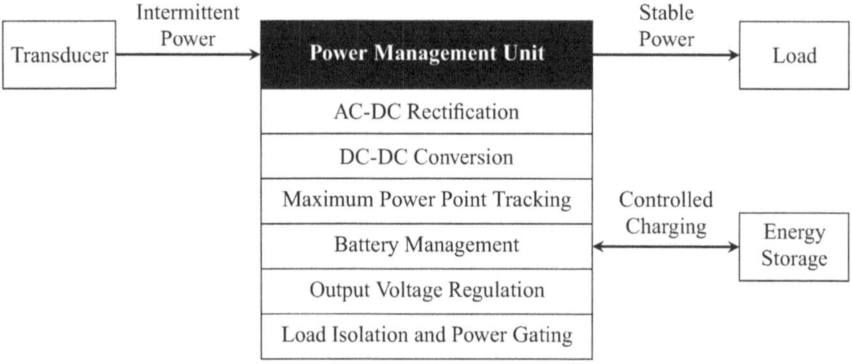

Fig. 19 Roles of power management unit in energy harvesting systems

Input Voltage Range	The allowable voltage range to keep the PMU operating, in particular the DC-DC boost converter. This must be compatible with the output voltage of the selected EH transducer. If the input drops below the minimum input voltage, the PMU would be switched off.
Cold Start Voltage	This refers to the minimum voltage to start the PMU from its power-off state. The cold start voltage is often significantly higher than the minimum input voltage. Designers need to ensure that the device is able to restart in this mode, in the event of temporary loss of the EH source.
Maximum Input Power	The maximum power generated by the selected EH transducer should not exceed this limit.
MPPT	If the MPPT setting of the PMU is fixed, it is necessary to make sure that it is close to the optimal point of the selected EH transducer. Otherwise, designers need calculate and re-program the MPPT setting accordingly, in order to maximize energy extraction from the transducer.
Quiescent Current	The current consumed by the PMU itself. It has to be taken into account in the calculation of the aforementioned power consumption profile.
Energy Storage Voltage	The PMU disconnects the storage if it is above or below the configured threshold voltages for safety purposes. These settings should be compatible with the energy storage in use to ensure that the storage is properly protected during operation.
Regulated Output	If the PMU does not provide the sufficient regulated output voltage and current for the consumption of processor, I/O peripherals and network interface, an additional voltage regulator might be necessary for the loads to draw power from the storage directly. The discharging of the energy storage must be monitored of to avoid safety issues, such as undervoltage and overcurrent.

4.3 Embedded Software Perspective

Self-sustainability can hardly be achieved with low-power hardware alone. Embedded software that runs on the devices also plays a critical role in facilitating their energy-neutral operations. Software-based power management strategies that associate with EH integration involve two areas in general, *energy prediction* and *load adaptation* [84].

Energy prediction through forecasting the future EH intake level is an essential part for devising power management strategies in EH-powered systems. Obviously, this is only applicable to predictable EH sources, mainly the Class 2 sources with quasi-cyclic patterns like solar and wind power. Through recognizing the current and future

Table 9 Examples of energy prediction algorithms for energy harvesting systems

Predictor	Brief description
Exponentially Weighted Moving Average (EWMA) [50]	Divides a harvesting cycle into multiple time slots, and compensates the unusual fluctuations to smoothen the estimated value by applying an EWMA filter. Historic data does not need to be stored to reduce memory usage
Weather Conditioned Moving Average (WCMA) [12]	Improves EWMA filtering by taking the result in the past few days into account, so as to compensate for weather-affected fluctuations
Machine learning of weather forecast [88, 89]	Applies machine learning to analyze and predict future energy availability based on weather forecast. This is not practical to be implemented on constrained devices
SunCast [60]	Provides a distribution of short-term sunlight prediction values. This is not practical to be implemented on constrained devices
Pro-energy [18]	Utilizes past observations to predict short-term (a few to 30 min) and medium-term (a few hours) availability

energy availability of both the device energy storage and EH source, the device can perform load adaptation to alter its operations to ensure energy neutrality throughout its lifetime. Several energy prediction algorithms surveyed from the literature are briefly described in Table 9.

Load adaptation techniques provide lifetime extension and performance maximization through altering a device's operating duty cycle (mainly for time-triggered devices). For example, when the detected energy level becomes low, the device will reduce its duty cycle to conserve energy, and increase it back again when sufficient energy is available. On the other hand, the device can also tune up the duty cycle to improve its performance [50] (e.g. to reduce the data measurement interval from one hour to 30 min), when the available energy is in excess. These techniques can generally be classified by two approaches, namely Energy-Reserve-Only and Harvest-Aware [84]. Energy-Reserve-Only load adaptation only considers the energy level in the storage, while the Harvest-Aware approach also takes the predicted EH availability into account to make more optimal adaptation decisions. Figure 20 shows some examples of load adaptation techniques that can be found in the literature.

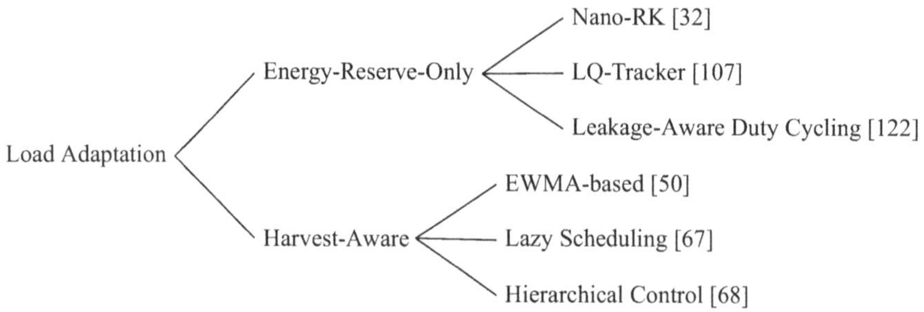

Fig. 20 Examples of load adaptation techniques for energy harvesting systems [84]

4.4 Summary of Design Considerations

The central idea of the proposed design workflow is that EH adoption in IoT devices requires to be considered at the early stage of the development life cycle. The key steps involving EH integration from the systems perspective include identifying possible EH sources in the deployment site, and minimizing the device power consumption by adopting low-power communication technologies and appropriate device operating modes. As for the device hardware perspective, using low-power hardware components and profiling the device's power consumption are the most crucial parts. Profiling power consumption helps size the EH transducer and energy storage appropriately, such that both self-sustainability goals and application requirements can be satisfied. From the embedded software perspective, EH systems may also require power management strategies, including energy prediction and load adaptation techniques introduced in Sect. 4.3, to further optimize their energy usage. Some major design considerations are summarized below, in terms of the four subsystems of the device architecture proposed in Sect. 2.1.

Processor	• Use low-power MCU with ultra-low power sleep modes. • Minimize the active duration per operating cycle to achieve the lowest possible duty cycle. • Avoid demanding computations on the device, and perform them in the cloud instead. • Adopt software-based power management strategies, such as energy prediction and load adaptation techniques, to maximize the device energy efficiency.
I/O	• Use low-power sensors and actuators. • Apply power-gating techniques to isolate I/O devices when they are unused.
Network	• Adopt low-power M2M communication technologies and energy-efficient network protocols whenever possible. • Use low-power network transceivers, and adjust their power levels appropriately to further reduce energy consumption. • Configure connection and data transmission timeout settings to prevent unexpected power drain during network failure.
Power	• Profile the power consumption of the device accurately. • Size EH transducer and energy storage appropriately to satisfy both device energy requirements and application constraints, such as physical dimensions and cost. • Include reasonable design margins during EH transducer and energy storage selection.

- Take both energy storage cycle life and degradation into account, and make sure that the storage can sustain the operations of the device safely throughout the designed lifetime.
- Select a suitable PMU and configure it properly to match the characteristics of the EH transducer in use.
- Ensure that the device can recover from scenarios of temporary loss of EH source.

5 Conclusions and Future Development

Internet of Everything (IoE) has become an inevitable technological trend, but powering the forthcoming billions of connected devices effectively remains a major challenge. Starting from the ultimate visions of IoE, this chapter firstly defines the distinguishing characteristics of Internet of Things (IoT), which is the vital pillar of IoE. The energy-related challenges of realizing IoT with distributed devices are examined through an extensive survey of machine-to-machine communication technologies and power supply options. Moreover, this chapter presents a comprehensive approach to analyze the energy consumption of IoT devices, and hence to illustrate the possibility for these devices to adopt energy harvesting (EH), which is deemed a prominent solution to tackle the energy-related challenges.

From concepts and principles to design considerations, this work attempts to bridge the gap between the state-of-the-art research and practical engineering needs, via establishing conceptual architectures and providing guidelines for designing EH-powered, self-sustaining and autonomous IoT devices. It can be foreseen that integrating EH technologies would be a pivotal research topic in the IoT arena, with regard to the demand of scaling up the coverage of devices to realize the envisioned future of IoE. Apart from enabling researchers and practitioners to apply the introduced techniques to IoT applications that make homes and cities smarter, it is hoped that this work lays the foundations for future research and development in two areas: Energy harvesting middleware and applications beyond Earth.

5.1 Energy Harvesting Middleware

The current approach of integrating EH in IoT devices requires significant additional effort in every aspect of the system, especially from the hardware and software perspectives, as discussed in Sect. 4. This inevitably incurs extra costs and complexity in designing and implementing EH-powered devices. It would be more desirable for developers to adopt EH, if a middleware layer is built into the commonly used IoT development frameworks and platforms for both devices and cloud servers. In the context of devices, middleware refers to the abstract layer that lies between hardware and the application logic [13], typically being the operating system that run on the microcontroller, such as Contiki [23] and RIOT [85]. Through integrating the aforementioned power management strategies into the operating system environment, developing EH-powered IoT devices can be accelerated with increased convenience. On the other hand, server-side EH middleware is also a possible new direction of development. Conventionally, the EH power management algorithms are computed solely on the embedded devices, which rendered more complex energy level and EH availability prediction impractical on the constrained devices. With the new IoT paradigm, a new approach can be adopted to shift such demanding computations to the cloud. Energy policies for individual IoT devices devised on the cloud can then be implemented on the devices through over-the-air update. Although some researchers have already endeavored to propose similar ideas for devices [24] and servers [89], further standardization and integration effort is needed to consolidate this concept systematically, and to ensure that the implementation works across heterogeneous IoT devices and platforms.

5.2 Applications Beyond Earth

Although the EH-powered IoT devices concerned in this chapter are purposed for various terrestrial applications, they resemble quite many design constraints and features of orbiting satellites, interplanetary probes and landers: Both have to be self-powered to operate autonomously for extended period of time, and post-deployment battery replacement and hardware repairs are impossible, despite the fact that spacecrafts need to survive in even more hostile radiation and thermal environments [34]. In preparation for the upcoming voyage to Mars and other destinations in the solar system, it is believed that low-cost autonomous sensors will play a vital role in scouting missions to survey the surface environment in advance of manned missions [102]. Self-powered, autonomous systems are also indispensable for picosatellites, also known as CubeSats [40], to reach out further in deep space. It is hoped that the proliferation of the EH-powered, tiny IoT devices can help improve lives on Earth, and also to foster the development of the counterparts that assist mankind to explore and settle in other new worlds in our universe.

References

1. Abdelwahab, S., B. Hamdaoui, M. Guizani, et al. 2014. Enabling smart cloud services through remote sensing: An internet of everything enabler. *IEEE Internet of Things Journal* 1(3): 276–288.
2. Akyildiz, I.F., W. Su, Y. Sankarasubramaniam, et al. 2002. Wireless sensor networks: A survey. *Computer Networks* 38(4): 393–422.
3. Al-Fuqaha, A., M. Guizani, M. Mohammadi, et al. 2015. Internet of Things: A survey on enabling technologies, protocols, and applications. *IEEE Communications Surveys & Tutorials* 17(4): 2347–2376.

4. Arms, S.W., et al. 2005. Power management for energy harvesting wireless sensors. In *Proceedings of SPIE*, ed. V.K. Varadan, San Diego.
5. Ashton, K. 2009. That 'Internet of Things' thing. http://www.rfidjournal.com/articles/view?4986. Accessed 9 Nov 2016.
6. Atmel Corporation. 2015. ATA8520D [datasheet]. http://www.atmel.com/Images/Atmel-9390-Smart-RF-ATA8520D_Datasheet.pdf. Accessed 30 Nov 2016.
7. Atmel Corporation. 2016. SAMB11 ultra low power BLE 4.1 SoC [datasheet]. http://www.atmel.com/Images/Atmel-42426-SmartConnect-SAMB11-SOC_Datasheet.pdf. Accessed 30 Nov 2016.
8. Atmel Corporation. 2016. SMART SAM R21 [datasheet]. http://www.atmel.com/Images/Atmel-42223. Accessed 30 Nov 2016.
9. Atzori, L., A. Iera, and G. Morabito. 2010. The Internet of Things: A survey. *Computer Networks* 54: 2787–2805.
10. Beeby, S., and N. White. 2010. *Energy harvesting for autonomous systems*. Norwood: Artech House.
11. Berger, A.S. 2002. *Embedded systems design: An introduction to processes, tools, and techniques*. Berkeley: CMP Books.
12. Bergonzini, C., D. Brunelli, and L. Benini. 2010. Comparison of energy intake prediction algorithms for systems powered by photovoltaic harvesters. *Microelectronics Journal* 41(11): 766–777.
13. Bischoff, U., and G. Kortuem. 2007. Life cycle support for sensor network applications. In *Proceedings of the 2nd International Workshop on Middleware for Sensor Networks (MidSens'07), November 2007*, 1–6. New York: ACM.
14. Bormann, C., M. Ersue., and A. Keranen. 2014. RFC 7228 - terminology for constrained-node networks. Internet Engineering Task Force. https://tools.ietf.org/html/rfc7228. Accessed 14 Nov 2016.
15. Brad, S. 2016. Greener datacenters for a brighter future: Microsoft's commitment to renewable energy. https://blogs.microsoft.com/on-the-issues/2016/05/19/greener-datacenters-brighter-future-microsofts-commitment-renewable-energy/. Accessed 19 Nov 2016.
16. Brill, J. 2015. How does the Amazon Dash Button work? http://www.forbes.com/sites/quora/2015/04/01/how-does-the-amazon-dash-button-work/#1d5b4f820280. Accessed 20 Nov 2016.
17. Buyya, R., A. Beloglazov, and J. Abawajy. 2010. Energy-efficient management of data center resources for cloud computing: A vision, architectural elements, and open challenges. arXiv:1006.0308. Accessed 20 Nov 2016.
18. Cammarano, A., et al. 2012. Pro-Energy: A novel energy prediction model for solar and wind energy-harvesting wireless sensor networks. In *2012 IEEE 9th International Conference on Mobile Ad Hoc and Sensor Systems (MASS 2012), October 2012*, 75–83. Las Vegas: IEEE.
19. Carvalho, C., and N. Paulino. 2014. On the feasibility of indoor light energy harvesting for wireless sensor networks. *Procedia Technology* 17: 343–350.
20. Centenaro, M., L. Vangelista, A. Zanella, et al. 2015. Long-range communications in unlicensed bands: The rising stars in the IoT and smart city scenarios. *IEEE Wireless Communications* 23(5): 60–67.
21. Chen, Y.-K. 2012. Challenges and opportunities of Internet of Things. In *17th Asia and South Pacific Design Automation Conference, January 2012*, 383–388. Sydney: IEEE.
22. Cheng, M.-Y. et al. 2013. Event-driven energy-harvesting wireless sensor network for structural health monitoring. In *38th Annual IEEE Conference on Local Computer Networks, October 2013*, 364–372. Sydney: IEEE.
23. Contiki. 2016. Contiki: The open source operating system for the Internet of Things. http://www.contiki-os.org. Accessed 19 Nov 2016.
24. DallOra, R., et al. 2014. SensEH: From simulation to deployment of energy harvesting wireless sensor networks. In *39th Annual IEEE Conference on Local Computer Networks Workshops, September 2014*, 566–573. Edmonton: IEEE.

25. Dargie, W., and C. Poellabauer. 2010. *Fundamentals of wireless sensor networks*. Chichester: Wiley.
26. Datta, S.K., et al. 2014. An IoT gateway centric architecture to provide novel M2M services. In *2014 IEEE World Forum on Internet of Things, March 2014*, 514–519. Seoul: IEEE.
27. De Sanctis, M., E. Cianca, G. Araniti, et al. 2016. Satellite communications supporting Internet of Remote Things. *IEEE Internet of Things Journal* 3(1): 113–123.
28. Dilhac, J.-M., and M. Bafleur. 2014. Energy harvesting in aeronautics for battery-free wireless sensor networks. *IEEE Aerospace and Electronic Systems Magazine* 29(8): 18–22.
29. European Technology Platform on Smart Systems Integration. 2008. Internet of Things in 2020: A roadmap for the future. http://www.smart-systems-integration.org/public/documents/publications/Internet-of-Things_in_2020_EC-EPoSS_Workshop_Report_2008_v3.pdf. Accessed 3 Oct 2016.
30. Ericsson. 2016. Cellular networks for massive IoT. https://www.ericsson.com/res/docs/whitepapers/wp_iot.pdf. Accessed 3 Dec 2016.
31. Espressif Systems. 2016. ESP32 datasheet. https://espressif.com/sites/default/files/documentation/esp32_datasheet_en.pdf. Accessed 30 Nov 2016.
32. Eswaran, A., et al. 2005. Nano-RK: An energy-aware resource-centric RTOS for sensor networks. In *26th IEEE International Real-Time Systems Symposium (RTSS'05), December 2005*, 256–265. Miami: IEEE.
33. Evans, D. 2012. The Internet of Everything - how more relevant and valuable connections will change the world. https://newsroom.cisco.com/video-content?type=webcontent&articleId=1111241. Accessed 9 Nov 2016.
34. Fortescue, P.W., J.P.W. Stark, and G. Swinerd (eds.). 2011. *Spacecraft Systems Engineering*, 4th ed. Chichester: Wiley.
35. Frank, R. 2013. *Understanding smart sensors*, 3rd ed. Norwood: Artech House.
36. Gomez, C., J. Oller, and J. Paradells. 2012. Overview and evaluation of Bluetooth Low Energy: An emerging low-power wireless technology. *Sensors* 12: 11734–11753.
37. Gozalvez, J. 2016. New 3GPP standard for IoT [mobile radio]. *IEEE Vehicular Technology Magazine* 11(1): 14–20.
38. Greenberg, A., J. Hamilton, D.A. Maltz, et al. 2008. The cost of a cloud. *ACM SIGCOMM Computer Communication Reviews* 39(1): 68–73.
39. Gubbi, J., R. Buyya, S. Marusic, et al. 2013. Internet of Things (IoT): A vision, architectural elements, and future directions. *Future Generation Computer Systems* 29(7): 1645–1660.
40. Heidt, H., J. Puig-Suari, A. Moore, et al. 2000. CubeSat: A new generation of picosatellite for education and industry low-cost space experimentation. http://digitalcommons.usu.edu/smallsat/2000/All2000/32/. Accessed 19 Nov 2016.
41. Hersent, O., D. Boswarthick, and O. Elloumi. 2011. *The Internet of Things: Key applications and protocols*. Chichester: Wiley.
42. Hu, Y., and V.O.K. Li. 2001. Satellite-based Internet: A tutorial. *IEEE Communications Magazine* 39: 154–162.
43. Karl, H., and A. Willig. 2007. *Protocols and architectures for wireless sensor networks*. Hoboken: Wiley.
44. Khatib, T., A. Mohamed, and K. Sopian. 2012. A review of solar energy modeling techniques. *Renewable and Sustainable Energy Reviews* 16(5): 2864–2869.
45. Hua, A.C.-C., and B.Z.-W. Syue. 2010. Charge and discharge characteristics of lead-acid battery and $LiFePO_4$ battery. In *The 2010 International Power Electronics Conference - ECCE ASIA -, June 2010*, 1478–1483. Sapporo: IEEE.
46. Iridium Communications Inc. 2016. Iridium 9603 transceiver. https://www.iridium.com/Products/Details/Iridium-9603?section=tech. Accessed 20 Nov 2016.
47. Jayakumar, H., et al. 2014. Powering the Internet of Things. In *ISLPED'14. Proceedings of the 2014 International Symposium on Low Power Electronics and Design*, 375–380. New York: ACM.
48. Jessen, J., et al. 2011. Design considerations for a universal smart energy module for energy harvesting in wireless sensor networks. In *2011 Proceedings of the 9th Workshop for Intelligent Solutions in Embedded Systems, July 2011*, 35–40. Regensburg: IEEE.

49. Jing, Q., A.V. Vasilakos, J. Wan, et al. 2014. Security of the Internet of Things: Perspectives and challenges. *Wireless Networks* 20(8): 2481–2501.
50. Kansal, A., J. Hsu, S. Zahedi, et al. 2007. Power management in energy harvesting sensor networks. *ACM Transactions on Embedded Computing Systems* 6: 32.
51. Kopetz, H. 1991. Event-triggered versus time-triggered real-time systems. In *Operating systems of the 90s and beyond, Dagstuhl Castle, July 1991. Lecture notes in computer science*, eds. Karshmer A, and Nehmer J, vol 563, 86–101. Berlin: Springer.
52. Kurs, A., A. Karalis, R. Moffatt, et al. 2007. Wireless power transfer via strongly coupled magnetic resonances. *Science* 317(5834): 83–86.
53. Lee, J.-S., et al. 2007. A comparative study of wireless protocols: Bluetooth, UWB, ZigBee, and Wi-Fi. In *IECON 2007–33rd Annual Conference of the IEEE Industrial Electronics Society, November 2007*, 46–51. Taipei: IEEE.
54. Lei, C.-U., K.L. Man, H.-N. Liang, et al. 2013. Building an intelligent laboratory environment via a cyber-physical system. *International Journal of Distributed Sensor Networks* 2013: 1–9.
55. Leonov, V. 2013. Thermoelectric energy harvesting of human body heat for wearable sensors. *IEEE Sensors Journal* 13(6): 2284–2291.
56. Lewandowski, S.M. 1998. Frameworks for component-based client/server computing. *ACM Computing Surveys* 30(1): 3–27.
57. Liang, N.-C., et al. 2006. Impact of node heterogeneity in ZigBee mesh network routing. In *2006 IEEE International Conference on Systems, Man and Cybernetics, October 2006*, 187–191. Taipei: IEEE.
58. Lien, S.-Y. 2014. Machine-to-machine communications: Technologies and challenges. *Ad Hoc Networks* 18: 3–23.
59. Loechte, A., F. Hoffmann, C. Krimphove, et al. 2014. Is $LiFePO_4$ technology ready for Internet of Things? *Advances in Internet of Things* 4(1): 1–4.
60. Lu, J., and K. Whitehouse. 2012. SunCast: Fine-grained prediction of natural sunlight levels for improved daylight harvesting. In *2012 ACM/IEEE 11th International Conference on Information Processing in Sensor Networks, April 2012*, 245–256. Beijing: IEEE.
61. Majumder, A., and J. Caffery. 2004. Power line communications: An overview. *IEEE Potentials* 23(4): 4–13.
62. McMahon, MM., and R. Rathburn. 2005. Measuring latency in Iridium satellite constellation data services. http://dodccrp.org/events/10th_ICCRTS/CD/papers/233.pdf. Accessed 4 Dec 2016.
63. Min, D., et al. 2012. Design and implementation of the multi-channel RS485 IoT gateway. In *2012 International Conference on Cyber Technology in Automation, Control, and Intelligent Systems, April 2012*, 366–370. Bangkok: IEEE.
64. Mineraud, J., O. Mazhelis, X. Su, et al. 2016. A gap analysis of Internet-of-Things platforms. *Computer Communications* 89: 5–16.
65. Minoli, D. 2013. *Building the internet of things with IPv6 and MIPv6: The evolving world of M2M communications*. Hoboken: Wiley.
66. Moschitta A., and I. Neri. 2014. Power consumption assessment in wireless sensor networks. http://www.intechopen.com/books/ict-energy-concepts-towards-zero-power-information-and-communication-technology/power-consumption-assessment-in-wireless-sensor-networks. Accessed 19 Nov 2016.
67. Moser, C., et al. 2007. Adaptive power management in energy harvesting systems. In *2007 Design, Automation and Test in Europe Conference and Exhibition, April 2007*, 1682. Nice: IEEE.
68. Moser, C., et al. 2008. Robust and low complexity rate control for solar powered sensors. In *2008 Design, Automation and Test in Europe, March 2008*, 230–235. Munich: IEEE.
69. Mosher, D. 2016. SpaceX just asked the FCC to launch 4,425 satellites. Business Insider. http://www.businessinsider.com/spacex-internet-satellite-constellation-2016-11. Accessed 4 Dec 2016.
70. Musiani, D., et al. 2007. Active sensing platform for wireless structural health monitoring. In *IPSN'07. Proceedings of the 6th International Conference on Information Processing in Sensor Networks, April 2007*, 390. New York: ACM.

71. Nokia. 2016. LTE evolution for IoT connectivity white paper. http://info.networks.nokia.com/LTE-M-Optimizing-LTE-for-the-Internet-of-Things-LP.html. Accessed 4 Dec 2016.
72. Nordic Semiconductor. 2014. nRF51822 product specification v3.3. http://infocenter.nordicsemi.com/pdf/nRF51822_PS_v3.3.pdf. Accessed 30 Nov 2016.
73. Nordman, B., and K. Christensen. 2013. Local power distribution with nanogrids. In *2013 International Green Computing Conference Proceedings, June 2013*, 1–8. Arlington: IEEE.
74. Park, C., and P. Chou. 2006. AmbiMax: Autonomous energy harvesting platform for multi-supply wireless sensor nodes. In *2006 3rd Annual IEEE Communications Society on Sensor and Ad Hoc Communications and Networks, September 2006*, 168–177. Reston: IEEE.
75. Parks, A.N., et al. 2013. A wireless sensing platform utilizing ambient RF energy. In *2013 IEEE Topical Conference on Biomedical Wireless Technologies, Networks, and Sensing Systems, January 2013*, 154–156. Austin: IEEE.
76. Patel, P., and D. Cassou. 2015. Enabling high-level application development for the Internet of Things. *Journal of Systems and Software* 103: 62–84.
77. Penella-Lpez, M.T., and M. Gasulla-Forner. 2011. *Powering autonomous sensors*. Dordrecht: Springer.
78. Porter, M.E., and J.E., Heppelmann. 2014. How smart, connected products are transforming competition. *Harvard Business Review*. https://hbr.org/2014/11/how-smart-connected-products-are-transforming-competition. Accessed 9 Nov 2016.
79. Priya, S., and D.J. Inman (eds.). 2009. *Energy harvesting technologies*. Boston: Springer.
80. Raghunathan, V., and P.H, Chou. 2006. Design and power management of energy harvesting embedded systems. In *ISLPED'06. Proceedings of the 2006 International Symposium on Low Power Electronics and Design, Tegernsee, October 2006*, 369. New York: ACM.
81. Raghunathan, V., et al. 2005. Design considerations for solar energy harvesting wireless embedded systems. In *IPSN'05. 4th International Symposium on Information Processing in Sensor Networks, April 2005*, 457–462. Los Angeles: IEEE.
82. Raza, U., P. Kulkarni, and M. Sooriyabandara. 2016 Low power wide area networks: A survey. arXiv:1606.07360. Accessed 19 Nov 2016.
83. Reddy, T. (ed.). 2010. *Lindens handbook of batteries*, 4th ed. New York: McGraw-Hill.
84. Renner, C., S. Unterschtz, V. Turau, et al. 2014. Perpetual data collection with energy-harvesting sensor networks. *ACM Transactions on Sensor Networks* 11(1): 1–45.
85. RIOT. 2016. RIOT–the friendly operating system for the Internet of Things. https://riot-os.org. Accessed 19 Nov 2016.
86. Roscoe, N.M., and M.D. Judd. 2013. Harvesting energy from magnetic fields to power condition monitoring sensors. *IEEE Sensors Journal* 13(6): 2263–2270.
87. Semtech Corporation. 2015. SX1272 datasheet. http://www.semtech.com/images/datasheet/sx1272.pdf. Accessed 30 Nov 2016.
88. Sharma, N., et al. 2010. Cloudy computing: Leveraging weather forecasts in energy harvesting sensor systems. In *2010 7th Annual IEEE Communications Society Conference on Sensor, Mesh and Ad Hoc Communications and Networks, June 2010*, 1–9. Boston: IEEE.
89. Sharma, N., et al. 2011. Predicting solar generation from weather forecasts using machine learning. In *2011 IEEE International Conference on Smart Grid Communications, October 2011*, 528–533. Brussels: IEEE.
90. Shnayder, V., et al. 2004. Simulating the power consumption of large-scale sensor network applications. In *SenSys'04. Proceedings of the 2nd International Conference on Embedded Networked Sensor Systems, November 2004*, 188. New York: ACM.
91. Silicon Laboratories. 2013. EM351/EM357 high-performance, integrated ZigBee/802.15.4 system-on-chip. http://www.silabs.com/Support%20Documents/TechnicalDocs/EM35x.pdf. Accessed 30 Nov 2016.
92. Stankovic, J.A. 2014. Research directions for the Internet of Things. *IEEE Internet of Things Journal* 1(1): 3–9.
93. Sudevalayam, S., and P. Kulkarni. 2011. Energy harvesting sensor nodes: Survey and implications. *IEEE Communications Surveys & Tutorials* 13(3): 443–461.

94. Takacs, A., et al. 2012. Energy harvesting for powering wireless sensor networks on-board geostationary broadcasting satellites. In *2012 IEEE International Conference on Green Computing and Communications, November 2012*, 637-640. Besancon: IEEE.
95. Texas Instruments. 2015. bq25570 nano power boost charger and buck converter for energy harvester powered applications (rev. E). http://www.ti.com/lit/ds/symlink/bq25570.pdf. Accessed 5 Dec 2016.
96. Texas Instruments. 2015. CC3200 SimpleLink Wi-Fi and Internet-of-Things solution, a single-chip wireless MCU (rev. F). http://www.ti.com/lit/ds/symlink/cc3200.pdf. Accessed 30 Nov 2016.
97. Texas Instruments. 2016. CC2630 SimpleLink 6LoWPAN, ZigBee wireless MCU (rev. B). http://www.ti.com/lit/ds/symlink/cc2630.pdf. Accessed 30 Nov 2016.
98. Texas Instruments. 2016. CC2640 SimpleLink Bluetooth wireless MCU (rev. B). http://www.ti.com/lit/ds/symlink/cc2640.pdf. Accessed 30 Nov 2016.
99. Torah, R., P. Glynne-Jones, M. Tudor, et al. 2008. Self-powered autonomous wireless sensor node using vibration energy harvesting. *Measurement Science and Technology* 19(12): 125202.
100. Torres, E.O., and G.A. Rincon-Mora. 2009. Electrostatic energy-harvesting and battery-charging CMOS system prototype. *IEEE Transactions on Circuits and Systems I: Regular Papers* 56(9): 1938–1948.
101. Tozlu, S., M. Senel, W. Mao, et al. 2012. Wi-Fi enabled sensors for Internet of Things: A practical approach. *IEEE Communications Magazine* 50(6): 134–143.
102. Ulmer, C., S. Yalamanchili and L. Alkalai. 2003. Wireless distributed sensor networks for in-situ exploration of Mars. http://citeseerx.ist.psu.edu/viewdoc/download?doi=10.1.1.116.324&rep=rep1&type=pdf. Accessed 7 Dec 2016.
103. Ungurean, I., et al. 2014. An IoT architecture for things from industrial environment. In *2014 10th International Conference on Communications, May 2014*, 1–4. Bucharest: IEEE.
104. u-blox. 2016. SARA-U2 series - data sheet. https://www.u-blox.com/sites/default/files/SARA-U2_DataSheet_(UBX-13005287).pdf. Accessed 30 Nov 2016.
105. u-blox. 2016. TOBY-L1 series - data sheet. https://www.u-blox.com/sites/default/files/products/documents/TOBY-L1_DataSheet_(UBX-13000868).pdf. Accessed 30 Nov 2016.
106. Van Son, V. 2013. Bringing new wind to Iowa. http://newsroom.fb.com/news/2013/11/bringing-new-wind-to-iowa/. Accessed 20 Nov 2016.
107. Vigorito, C.M., et al. 2007. Adaptive control of duty cycling in energy-harvesting wireless sensor networks. In *2007 4th Annual IEEE Communications Society Conference on Sensor, Mesh and Ad Hoc Communications and Networks, June 2007*, 21–30. San Diego: IEEE.
108. Volakis, J.L., U. Olgun, and C.-C. Chen. 2012. Design of an efficient ambient WiFi energy harvesting system. *IET Microwaves, Antennas and Propagation* 6(11): 1200–1206.
109. Vullers, R., R. Schaijk, H. Visser, et al. 2010. Energy harvesting for autonomous wireless sensor networks. *IEEE Solid-State Circuits Magazine* 2(2): 29–38.
110. Wang, Q., et al. 2006. A realistic power consumption model for wireless sensor network devices. In *2006 3rd Annual IEEE Communications Society on Sensor and Ad Hoc Communications and Networks, September 2006*, 286–295. Reston: IEEE.
111. Wang, W., V. Cionca, N. Wang, et al. 2013. Thermoelectric energy harvesting for building energy management wireless sensor networks. *International Journal of Distributed Sensor Networks* 2013: 1–14.
112. Wang, Y., P. He, H. Zhou, et al. 2011. Olivine $LiFePO_4$: Development and future. *Energy and Environmental Science* 4(3): 805–817.
113. Wang, Y.-P., X., Lin, A. Adhikary, et al. 2016. A primer on 3GPP Narrowband Internet of Things (NB-IoT). arXiv:1606.04171. Accessed 20 Nov 2016.
114. Welbourne, E., L. Battle, G. Cole, et al. 2009. Building the Internet of Things using RFID: The RFID ecosystem experience. *IEEE Internet Computing* 13(3): 48–55.
115. Wolf, M. 2016. *Computers as components: Principles of embedded computing system design*, 4th ed. Burlington: Morgan Kaufmann.

116. Xie, L., S. Yi, Y.T. Hou, et al. 2013. Wireless power transfer and applications to sensor networks. *IEEE Wireless Communications* 20(4): 140–145.
117. Xu, B., L. Xu, H. Cai, et al. 2014. Ubiquitous data accessing method in IoT-based information system for emergency medical services. *IEEE Transactions on Industrial Informatics* 10(2): 1578–1586.
118. Xu, L.D., W. He, and S. Li. 2014. Internet of Things in industries: A survey. *IEEE Transactions on Industrial Informatics* 10(4): 2233–2243.
119. Yamada, A., S.C. Chung, and K. Hinokuma. 2001. Optimized $LiFePO_4$ for lithium battery cathodes. *Journal of Electrochemistry Society* 148(3): A224.
120. Yu, H., and Q. Yue. 2012. Indoor light energy harvesting system for energy-aware wireless sensor node. *Energy Procedia* 16: 1027–1032.
121. Yuan, S., Y. Huang, J. Zhou, et al. 2015. Magnetic field energy harvesting under overhead power lines. *IEEE Transactions on Power Electronics* 30(11): 6191–6202.
122. Zhu, T., et al. 2009. Leakage-aware energy synchronization for wireless sensor networks. In *MobiSys'09. Proceedings of the 7th International Conference on Mobile Systems, Applications, and Services, June 2009*, 319. New York: ACM.

A Detailed Analysis of IoT Platform Architectures: Concepts, Similarities, and Differences

Jasmin Guth, Uwe Breitenbücher, Michael Falkenthal, Paul Fremantle, Oliver Kopp, Frank Leymann and Lukas Reinfurt

Abstract The IoT is gaining increasing attention. The overall aim is to interconnect the physical with the digital world. Therefore, the physical world is measured by sensors and translated into processible data, and data has to be translated into commands to be executed by actuators. Due to the growing interest in IoT, the number of platforms designed to support IoT has risen considerably. As a result of different approaches, standards, and use cases, there is a wide variety and heterogeneity of IoT platforms. This leads to difficulties in comprehending, selecting, and using appropriate platforms. In this work, we tackle these issues by conducting a detailed analysis of several state-of-the-art IoT platforms in order to foster the understanding of the (i) underlying concepts, (ii) similarities, and (iii) differences between them. We show that the various components of the different platforms can be mapped to an abstract reference architecture, and analyze the effectiveness of this mapping.

J. Guth (✉) · U. Breitenbücher · M. Falkenthal · F. Leymann · L. Reinfurt
Institute of Architecture of Application Systems, University of Stuttgart, Stuttgart, Germany
e-mail: guth@iaas.uni-stuttgart.de

U. Breitenbücher
e-mail: breitenbuecher@iaas.uni-stuttgart.de

M. Falkenthal
e-mail: falkenthal@iaas.uni-stuttgart.de

F. Leymann
e-mail: leymann@iaas.uni-stuttgart.de

L. Reinfurt
e-mail: reinfurt@iaas.uni-stuttgart.de

P. Fremantle
School of Computing, University of Portsmouth, Portsmouth, UK
e-mail: paul.fremantle@port.ac.uk

O. Kopp
Institute for Parallel and Distributed Systems, University of Stuttgart, Stuttgart, Germany
e-mail: kopp@ipvs.uni-stuttgart.de

B. Di Martino et al. (eds.), *Internet of Everything*, Internet of Things,
https://doi.org/10.1007/978-981-10-5861-5_4

1 Introduction

The vision of the Internet of Things (IoT) describes a future where many everyday objects are interconnected through a global network. They collect and share data of themselves and their surroundings to allow widespread monitoring, analyzation, optimization, and control [27]. Until recently this was merely a vision, but in recent years this has slowly developed into a reality. Ever decreasing prices, dimensions, and energy requirements of electronics now allow tiny devices to unobtrusively measure their surroundings. Many devices use low-energy communication technology to send those measurements to other, more powerful components, such as bluetooth gateways, mobile phones, or WiFi hotspots. Devices are increasingly incorporating long-range wireless technologies such as LoRa[1] or existing 2G and 3G networks. Local edge processors, hubs, or internet services in turn analyze and process IoT sensor data to create new knowledge, which can be used to act back on the environment through actuators. In short, the IoT can be seen as a giant cyber-physical control loop. In that context, the term "Machine-to-Machine communication" (M2M [9]) is often used to describe such a setting.

Different incarnations of IoT systems for varying use cases have been created over the years by companies and research institutions. Smart Homes are one example of such IoT systems [30]. In other areas, similar developments are underway, such as Connected Cars [20], Smart Cities [31], Demand Side Management [22], Smart Grids [11], or Smart Factory systems [24].

While local processing of the data generated by these systems is possible and a reasonable approach for use cases where low latency is required, cloud based platforms are used for processing and analyzing larger data sets [7]. As a result, over one hundred [29] such platforms have been created over the last few years. Some examples include AWS IoT,[2] FIWARE,[3] OpenMTC,[4] and SmartThings[5].

These platforms come in various shapes and sizes. While standardization efforts are ongoing, there are no generally agreed-on standards for IoT at this time [8]. Rather, development of these platforms has often taken place in silos [38]. These different environments have influenced not only the choice of concepts and technology, but also the choice of terminology. As a result, the platform landscape has become very heterogeneous. At the same time, however, all these solution do roughly the same things: they allow connecting different devices, accessing and processing their data, and using the knowledge gained through this activity to create automated control.

The heterogeneity of these approaches creates an issue for someone who has to select one of these solutions. Finding the right platform for a use case becomes very time consuming when each solution uses different technologies and terminology. You have to read and understand the descriptions and documentation of each platform to

[1]https://www.lora-alliance.org/.

[2]https://aws.amazon.com/en/iot/.

[3]https://www.fiware.org/.

[4]http://www.openmtc.org/.

[5]http://www.smartthings.com/.

make a decision. This requires not only time, but also the technical knowledge to be able to understand and compare the different concepts.

A reference architecture which maps to existing architecture descriptions and offers a unified terminology helps in this selection process. Not only does it allow for easier comparison between platforms, but it also provides a useful framework as a starting point for new developments. Therefore, we define in the next section an IoT reference architecture based on existing platforms. The IoT reference architecture is kept purposely abstract to make it applicable in a wide range of situations. It was first introduced in our former work by Guth et al. [17]. The work at hand extends that previous work by a refinement of the description, the extension of the analysis to eight IoT platforms, and a more extensive survey on related work.

The remainder of this paper is structured as follows: In Sect. 2 we present our reference architecture. We describe the different components and the possible ways they communicate. In Sect. 3 we compare our architecture against eight existing platforms to show that it is generally applicable. To deepen and clarify the differences between the considered platforms we extended our previous work by describing the mapping of our reference architecture onto more IoT solutions. We also provide a summarized comparison, illustrated within Table 1. In Sect. 4 we investigate the differences of our reference architecture to other existing approaches. Finally, we summarize, conclude, and outline possible future work in Sect. 5.

2 Reference Architecture

This section presents an IoT reference architecture (Fig. 1) which (i) offers a unified terminology and (ii) maps to existing architecture descriptions. We start by defining all components shown in Fig. 1 starting from the bottom. To clearly distinguish between the concepts presented in this work and similar or equal concepts presented in the considered platforms and related work, we accentuate the elements of the IoT reference architecture presented in this work by italics.

2.1 Sensor

A *Sensor* is a hardware component which captures information on the physical environment by "respond[ing] to a physical stimulus (as heat, light, sound, pressure, magnetism, or a particular motion)" [25]. For instance, by measuring the humidity within a room, a *Sensor* positioned within that room captures the humidity level of the room. *Sensors* transmit the captured information using electrical signals to *Devices*, to which they are connected to. This connection can be established (i) by wire or (ii) wirelessly. Wired connection includes an integration of *Sensors* into a *Device*. A *Sensor* may be configured using software, but cannot run software by itself.

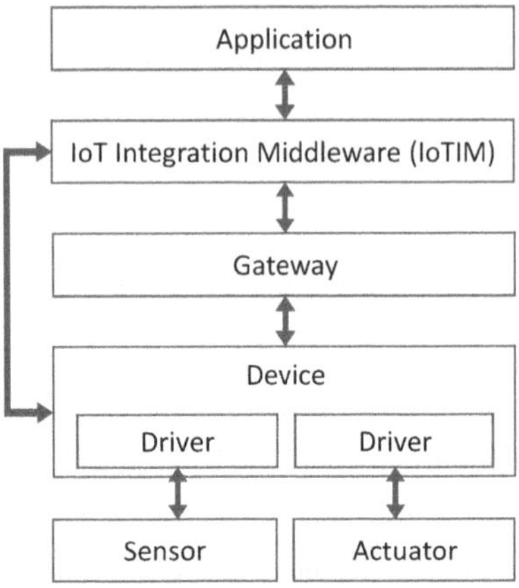

Fig. 1 IoT reference architecture based on [17]

2.2 Actuator

An *Actuator* is a hardware component which manipulates the physical environment. *Actuators* receive commands from their connected *Device* and translate these electrical signals into some kind of physical action. For instance, an *Actuator* turning on or off a ventilation within a room acts on the physical environment by influencing the humidity of the room. Similar to *Sensors*, the connection to the device can be established (i) by wire or (ii) wirelessly, whereby a wired connection includes the integration into a *Device*. Furthermore, an *Actuator* may be configured using software, but cannot run software by itself.

2.3 Device

A *Device* is a hardware component which (i) is connected to *Sensors* and/or *Actuators* by wire or wirelessly or (ii) even integrates these components. *Devices* have a processor and storage capacity to run software and to establish a connection to the *IoT Integration Middleware*. For instance, the outdoor module of the Netatmo weather station[6] represents a *Device* with integrated *Sensors*. Thus, *Devices* are the entry point of the physical environment to the digital world. A *Driver* is software running on the *Device* enabling uniform access to heterogeneous *Sensors* and *Actuators*. *Devices* are either (i) self-contained or (ii) connected to another, bigger system. The *IoT Integration Middleware* represents such a system.

[6]https://www.netatmo.com/product/weather/weatherstation.

2.4 Gateway

In case a *Device* is not capable of directly connecting to further systems, it is connected to a *Gateway*. A *Gateway* provides required technologies and mechanisms to translate between different protocols, communication technologies, and payload formats. It forwards communication between *Devices* and further systems. For instance, the indoor module of the Netatmo weather station (see footnote 6) is a *Device* with integrated *Sensors*, acting as a *Gateway* for the outdoor module of the Netatmo weather station. When the *Gateway* receives a message in a proprietary binary format from the *Device*, it translates the binary format into a more common format, such as JSON, and forwards the data to the intended system over IP, for example. If necessary, the *Gateway* may likewise translate commands sent from systems to *Devices* into communication technologies, protocols, and formats supported by the respective *Device*.

2.5 IoT Integration Middleware

The *IoT Integration Middleware* (IoTIM) serves as an integration layer for different kinds of *Sensors*, *Actuators*, *Devices*, and *Applications*. It is responsible for (i) receiving data from the connected *Devices*, (ii) processing the received data, (iii) providing the received data to connected *Applications*, and (iv) controlling *Devices*. An example for processing is to evaluate condition-action rules and sending commands to *Actuators* based on this evaluation. A *Device* can communicate directly with the *IoT Integration Middleware* if it supports an appropriate communication technology, such as IP over Ethernet or WiFi, a corresponding transport protocol, such as HTTP or MQTT, and a compatible payload format, like, e.g., JSON. Otherwise, the *Device* communicates over a *Gateway* with the *IoT Integration Middleware*. The *IoT Integration Middleware* is not limited to the functionality described above. It may comprise all kinds of functionality that are required by a certain cyber-physical system, for instance, a time-series database or graphical dashboards. Additionally, the management of *Devices* and users as well as the aggregation and utilization of received data may be performed. For instance, the SmartThings[5] platform can be used with the Netatmo *Devices*. Typically, an *IoT Integration Middleware* can be accessed using APIs, for example, HTTP-based REST APIs.

2.6 Application

The *Application* component represents software which uses the *IoT Integration Middleware* (i) to gain insight into the physical environment and/or (ii) to manipulate the physical world. It does so by requesting *Sensor* data or by controlling physical actions using *Actuators*. For instance, a software system that controls the temperature of a building represents an *Application* connected to an *IoT Integration Middleware*. An *Application* can also be another *IoT Integration Middleware*.

2.7 Summary

This section presented the reference architecture consisting of six component types. When implementing the architecture, components can be omitted. This might be the case, if the platform is only used to measure changes of the physical world. For instance, a platform gathering the CO2 level within the air, may have no *Actuators* connected, if the system is only used to measure and collect the data. Another example for omitted components are platforms with connected *Devices* capable of the required technologies to communicate directly with the *IoT Integration Middleware*, so no *Gateway* is needed for an appropriate message exchange.

3 Comparison of IoT Platforms

In this section, the IoT reference architecture is mapped onto four open-source platforms and four proprietary IoT solutions. The selected open-source platforms are FIWARE (see footnote 3), OpenMTC,[7] SiteWhere,[8] and Webinos.[9] The proprietary solutions are AWS IoT (see footnote 2), IBM's Watson IoT Platform,[10] Microsoft's Azure IoT Hub,[11] and Samsung's SmartThings (see footnote 5). During the comparison the component's functionality and their naming are the key areas that are compared with the reference architecture. The comparison of each platform will be described in detail. Additionally, a composite comparison is presented, where the main aspects are discussed. This section extends our previous analysis presented by Guth et al. [17] by a detailed analysis of four more platforms: Webinos, IBM's Watson IoT Platform, Microsoft's Azure IoT Hub, and Samsung's SmartThings. Moreover, we refined our previous analysis of FIWARE, OpenMTC, SiteWhere, and AWS IoT [17] in terms of a more detailed reference architecture mapping. In addition, we added a detailed comparison table of the platforms in Sect. 3.9.

3.1 FIWARE

The architecture and the mapping of the open-source platform FIWARE (see footnote 3) onto our IoT reference architecture is shown in Fig. 2. FIWARE is funded by the European Union and the European Commission. It provides an enhanced OpenStack-based[12] cloud, where its capabilities and Catalogue is hosted. The FIWARE Cata-

[7]http://www.open-mtc.org/.

[8]http://www.sitewhere.org/.

[9]http://www.webinos.org/.

[10]http://www.ibm.com/internet-of-things/.

[11]https://azure.microsoft.com/de-de/suites/iot-suite/.

[12]https://www.openstack.org/.

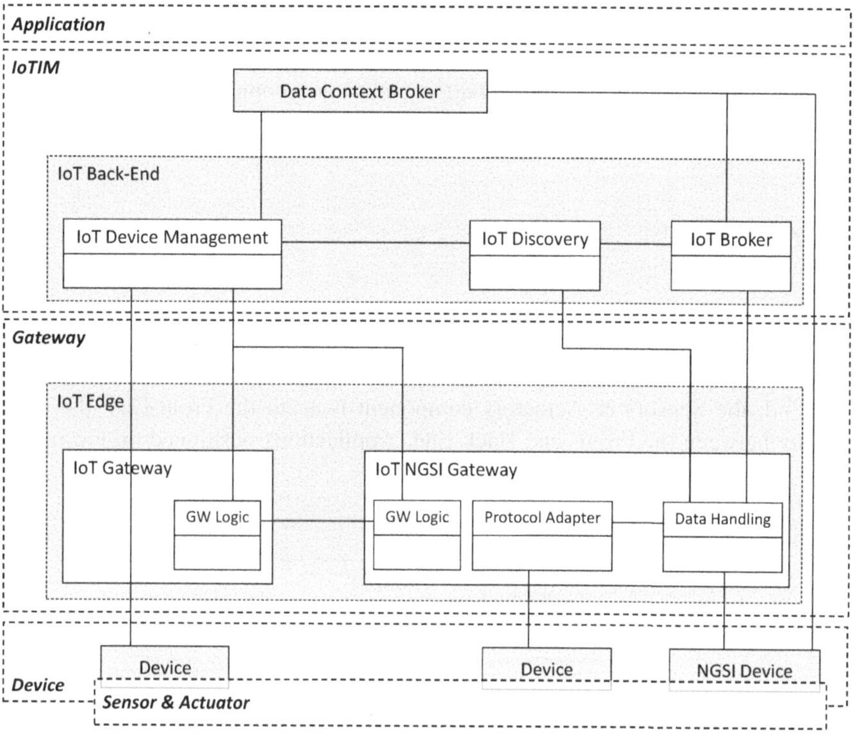

Fig. 2 FIWARE architecture based on [12]

logue contains Generic Enablers (GEs), representing a rich library of components. Within the architecture in Fig. 2, only the GEs of the IoT part are shown. FIWARE only distinguishes between Devices and NGSI[13] Devices. Since the FIWARE documentation describes that devices may have integrated sensors and actuators, all device components are comprised by our *Device*, *Sensor*, and *Actuator* components. Devices can communicate directly with the IoT Back-End or via a Gateway, which is positioned within the IoT Edge. Both the IoT Gateway and the IoT NGSI Gateway enable and manage the communication of devices with the IoT Back-End. Consequently, the IoT Edge represents the *Gateway* of our IoT reference architecture. The IoT Back-End and the Data Context Broker provide the main functionality of FIWARE, and therefore they are encapsulated by our *IoT Integration Middleware*. The documentation of FIWARE describes how further applications can be connected through the Data Context Broker to the platform. Although the application component is not represented in the FIWARE architecture diagram, based on this description we therefore position our *Application* component on top of the Data Context Broker.

[13]Next Generation Service Interfaces [28].

In regards to FIWARE, our IoT reference architecture covers each component of the architecture. As described above, the definition of the device component differs from ours and, hence, the *Sensor*, *Actuator*, and *Device* components partly overlap. Nevertheless, there is still an appropriate mapping to our definition.

3.2 OpenMTC

Figure 3 shows the architecture of OpenMTC (see footnote 7), which is an open-source, cloud-enabled IoT platform, and its comparison against our IoT reference architecture. OpenMTC is composed of the following components: The Front- and Back-End, the Sensors & Actuators component beneath the Front-End, the connectivity between the Front- and Back-End, Applications positioned on top of the

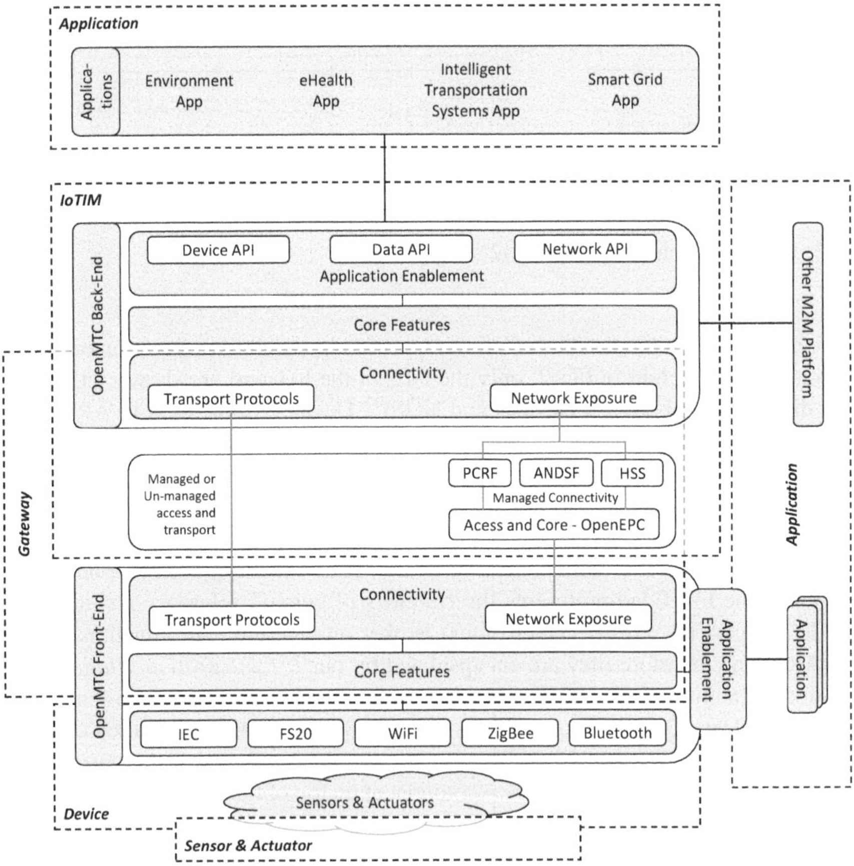

Fig. 3 OpenMTC architecture based on [13]

Back-End and on the right side of the Front-End, as well as a component to connect Other M2M Platforms to the Back-End. Obviously, the Sensors & Actuators component represents our *Sensors* and *Actuators*. The documentation of OpenMTC further describes that the Sensors & Actuators components along with the lowest level of the Front-End, which enables communication, represent *Devices* equivalent to our reference architecture. The Core Features and Connectivity components of the Front-End, the connectivity between the Front- and Back-End, as well as the Connectivity component of the Back-End provide all required functionality to enable the communication between a device and the middleware, such as message translation. Consequently, those components represent our *Gateway*. Since the OpenEPC component (offering connectivity between the Front- and Back-End) provides further functionality, such as applying rules and filtering, it is encompassed by the *IoT Integration Middleware* as well. Additionally, the *IoT Integration Middleware* encapsulates the Connectivity, Core Features, and Application Enablement components of the OpenMTC Back-End. Those components provide the main functionality of the platform. The Application Enablement components provide all functionality to connect further applications to the middleware. Hence, the Applications and Other M2M Platform components are covered by the *Applications* component of our reference architecture.

Considering OpenMTC, its architecture can be mapped onto our IoT reference architecture. Nevertheless, the *Device*, *Sensor*, and *Actuator* components, as well as the *Gateway* and *IoT Integration Middleware* are partly overlapping, which is still appropriate to the definition of our reference architecture.

3.3 SiteWhere

Figure 4 shows the architecture and the mapping of the open-source IoT platform SiteWhere (see footnote 8) onto our reference architecture. The Data from Devices and Commands to Devices components are comprised by the *Device* component of our reference architecture. Furthermore, they also represent our *Sensor* and *Actuator* components, since they are not explicitly depicted within the architecture. As the devices can communicate via diverse protocols with the platform, the concept of a *Gateway* is present between the devices and the platform [32], but not pictured as a separate component. The main functionality of the platform is provided by the SiteWhereTenant Engine, including the Device Management and the Communication Engine. Consequently, those components are comprised by the *IoT Integration Middleware* of our reference architecture. The REST APIs and Integration component enables the connection of further *Applications* to the platform.

Considering the architecture of SiteWhere, each component of our IoT reference architecture is represented. Even though, some components are overlapping, our definition of the IoT reference architecture components still holds.

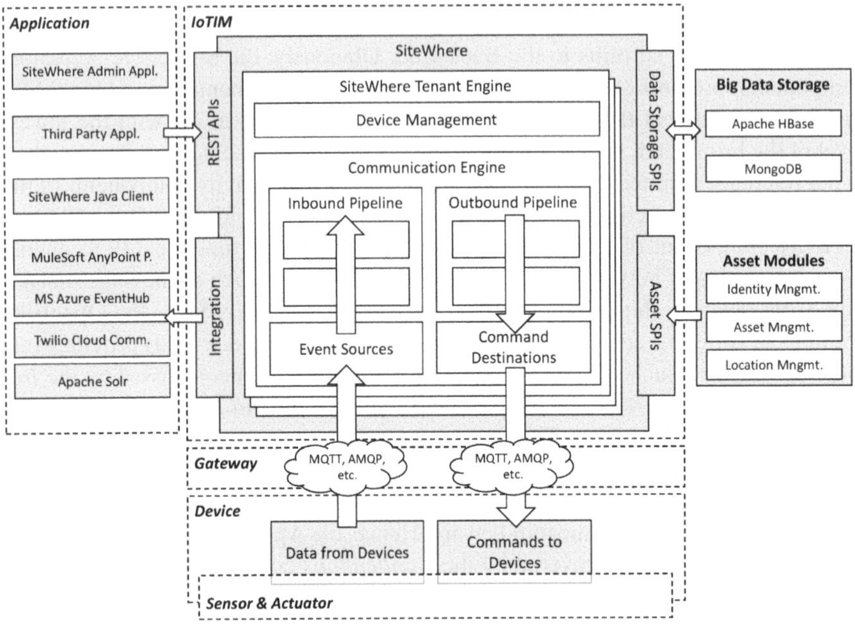

Fig. 4 SiteWhere architecture based on [32]

3.4 Webinos

The Webinos (see footnote 9) middleware and architecture is an open-source middleware for the IoT and mobile devices, sponsored by the European Union FP7 project. The aim of Webinos is to provide a secure framework for personal devices to communicate and for individuals to publish data to third-parties and to other individuals. As such it takes a different approach to an IoT platform by being centered around the person. The mapping of the Webinos approach onto our IoT reference architecture is shown within Fig. 5. The main components of the Webinos architecture are the Personal Zone Hub (PZH), and the Personal Zone Proxy (PZP). The PZH provides the *Gateway*, where each *Device* connects to. The PZH also provides local communications between devices by acting as a messaging hub. In this regard it performs the functions of our *IoT Integration Middleware*. The PZH does not inherently support any *Applications* to run locally, but it provides the APIs that allow third-party applications to be built and to communicate with devices, which is a core function of the *IoT Integration Middleware* layer in our architecture. Each device runs a local component, the PZP, that aggregates sensor data and actuator commands and communicates with the PZH. One unique aspect of Webinos is that when multiple PZPs (on multiple devices) have connected to the same PZH, they can then communicate in a peer-to-peer fashion. The PZP and PZH sync to allow this to happen.

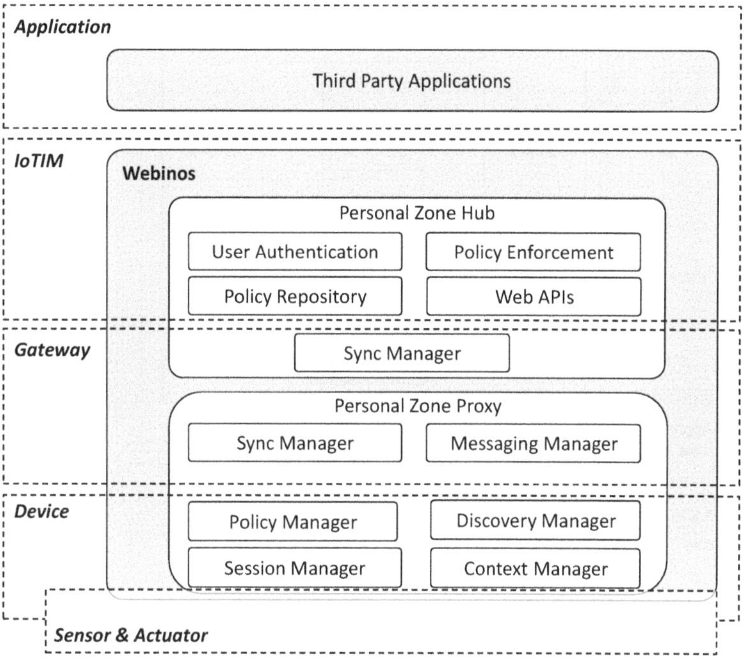

Fig. 5 Webinos architecture based on [15, 34]

Each component of our IoT reference architecture is represented in the Webinos system. Because the PZH and PZP overlap in function, the separation into our reference architecture is more complicated. However, there is a clear mapping that certain components are performing *Gateway* and *IoT Integration Middleware* functions, as depicted in the diagram. As an example, because the PZPs can communicate in a peer-to-peer fashion without the PZH acting as an intermediary, we must assign some of the *Gateway* functionality to the PZP. Each PZP is deployed on a *Device*. Once again, the Webinos system does not explicitly call out the difference between *Devices*, *Sensors*, and *Actuators*, but there is full support for both sensors and actuators in Webinos and, therefore, it supports the reference architecture.

While the reference architecture does not explicitly deal with the person-centered approach of Webinos, we can clearly map each person's Webinos system to an individual instance of the reference architecture.

3.5 AWS IoT

Figure 6 shows the architecture of Amazon Web Services IoT (see footnote 2) (AWS IoT) and its mapping onto our reference architecture. AWS IoT is a managed cloud platform for the IoT, pursuing the concept of Things instead of devices. Since AWS

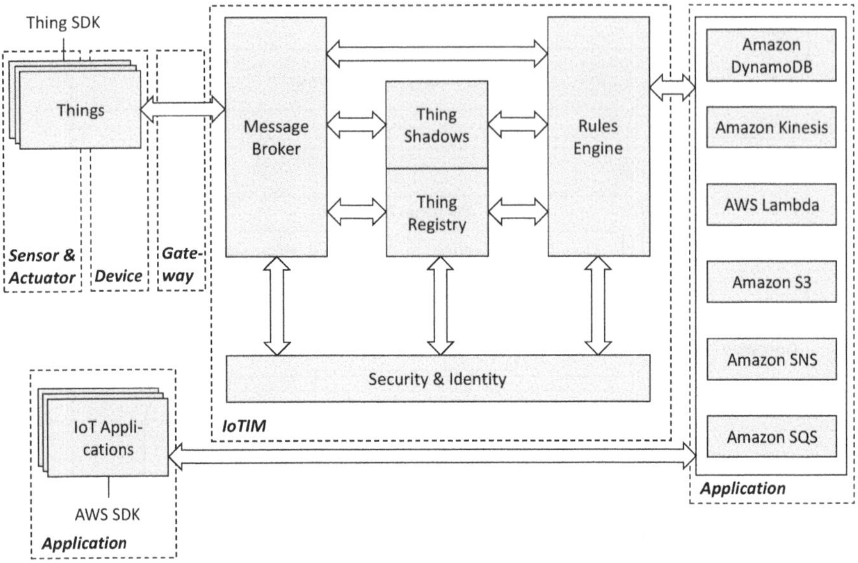

Fig. 6 AWS IoT architecture based on [2]

uses Things synonymous to devices with integrated sensors and actuators, the Things component is covered by our *Device*, *Sensor*, and *Actuator* components. Furthermore, a *Gateway* component is not represented within the architecture, but following the documentation [2], it is located between the Things and Message Broker components. The Message Broker, Thing Shadows, Thing Registry, Rules Engine, and the Security & Identity components provide the main functionality of the platform. Hence, they represent the *IoT Integration Middleware* component of our IoT reference architecture. Our *Application* component encapsulates the already integrated data processing services, such as AWS Lambda or Amazon Kinesis, and, additionally, the IoT Applications component, which enables the connection of further applications.

Considering AWS IoT, each component of our IoT reference architecture is represented. Even though AWS follows a different concept of devices, it is still appropriate to our definition of the components.

3.6 IBM Watson IoT Platform

Figure 7 shows the architecture of IBM's cloud-based Watson IoT Platform (see footnote 10). Within the figure, *Devices*, *Sensors*, and *Actuators* are not represented. Since the Connect component takes care of the connection of *Devices* to the platform, the *Device*, *Sensor*, and *Actuator* components of our terminology partly cover the Connect component. Furthermore, the Connect component is responsible

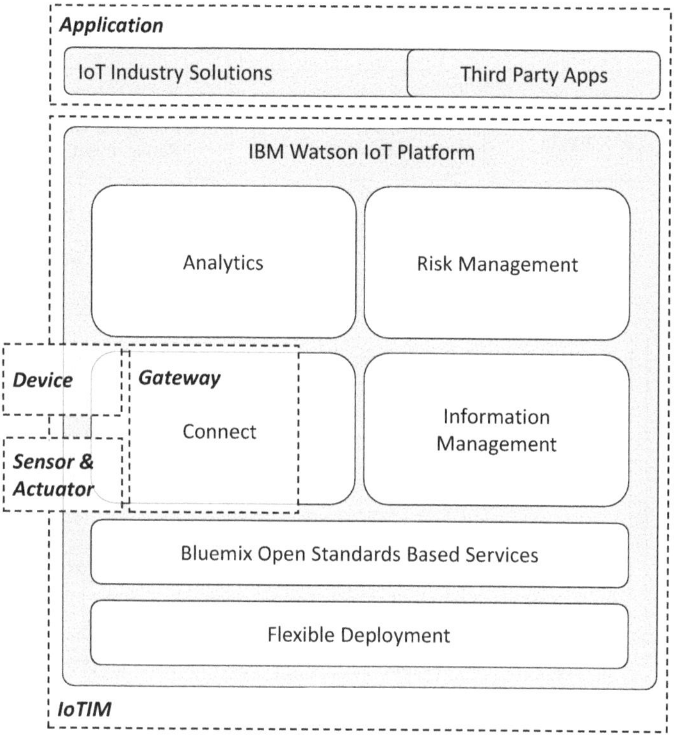

Fig. 7 IBM Watson IoT Platform architecture based on [19]

for a corresponding message translation, hence, it is encompassed by our *Gateway* component. The Connect component also provides further event handling functionality, thus, it is covered by our *IoT Integration Middleware* as well. In addition, the Analytics, Risk Management, and Information Management components provide the core functionality of the platform, thus, they are covered by the *IoT Integration Middleware* of our terminology. The Bluemix Open Standards Based Services component and the Flexible Deployment component build the basis of the platform. The IoT Industry Solutions and Third Party Apps components are encompassed by our *Application* component, since they enable the connection of further applications.

With consideration to the IBM Watson IoT Platform, our IoT terminology can be mapped onto it. Even though the *Device*, *Sensor*, *Actuator*, and *Gateway* components are not represented in particular, they are part of the Connect component.

3.7 *Microsoft Azure IoT Hub*

The Azure IoT Hub (see footnote 11) is a managed, cloud-based service, provided by Microsoft. Its architecture and the mapping onto our IoT reference architecture

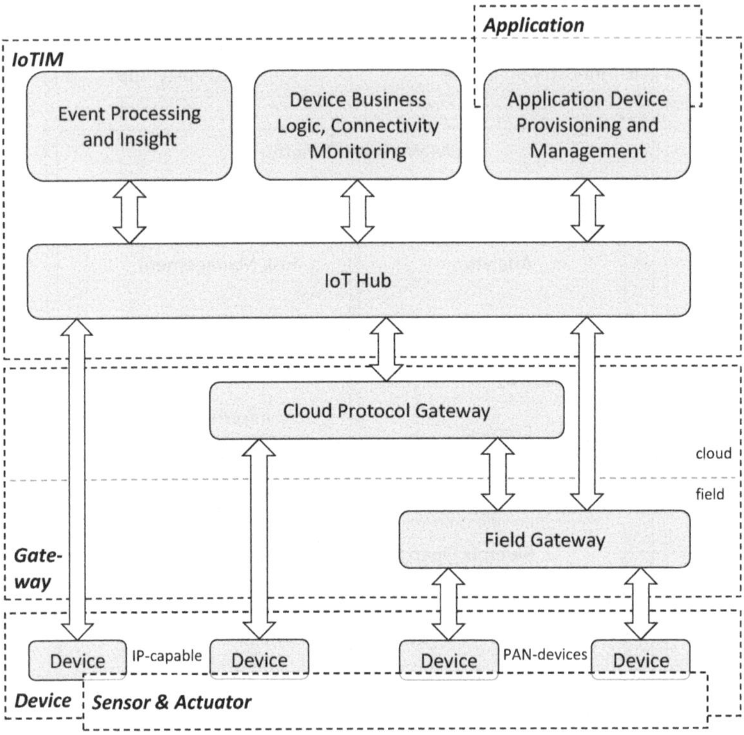

Fig. 8 Microsoft Azure IoT Hub architecture based on [6]

is shown in Fig. 8. The main component is the IoT Hub, where all remaining components are connected to. Since Microsoft only separates between IP-capable and Personal Area Network (PAN) Devices, those map to our *Device*, *Sensor*, and *Actuator* components. IP-capable devices may communicate directly or via a Cloud Protocol Gateway with the IoT Hub, whereby PAN Devices additionally need a Field Gateway to perform local management services, such as managing access and information flow. Hence, the Cloud Protocol and the Field Gateway are covered by our *Gateway* component. The core functionality of the solution is provided by the IoT Hub, the Event Processing and Insight, the Device Business Logic, Connectivity Monitoring, and the Application Device Provisioning and Management components, hence, they are covered by our *IoT Integration Middleware*. Furthermore, the Application Device Provisioning and Management component also enables the connection of further *Applications*.

With regard to the Microsoft Azure IoT Hub, each component of our IoT reference architecture is represented. Some components are overlapping and, again, it is not further distinguished between *Devices*, *Sensors*, and *Actuators*, but this is still appropriate to the definition of our IoT reference architecture components.

3.8 *SmartThings*

SmartThings (see footnote 5) is an IoT platform provided by Samsung for a smart home environment. The architecture and the corresponding mapping with our IoT reference architecture is shown in Fig. 9. It is composed of three core elements, namely the Device Type Handlers, the Subscription Processing, and the Application Management System including the SmartApp Management and Execution, where all further components are connected to. SmartThings combines Sensors, Actuators, Devices, Users, and Things within one component. They further distinguish between this composed component and another Clients (-Devices) component. Since Clients (-Devices) may also contain *Sensors* and *Actuators*, both components are covered by our *Device* component, with the *Sensor* and the *Actuator* components partly overlapping. Since the Device Type Handlers translate event-messages into a normalized SmartThings event, and the Hub Connectivity and the Client Connectivity enables the connection of Devices to the platform, those components represent the *Gateway*. The core functionality of the platform is provided by the Subscription Processing and the Application Management System, including the SmartApp Management & Execution components. Consequently, they cover the *IoT Integration Middleware* of our reference architecture. The Event Stream, Web UI, Core APIs, External System, and Physical Graph components represent possibly connected *Applications* to the platform.

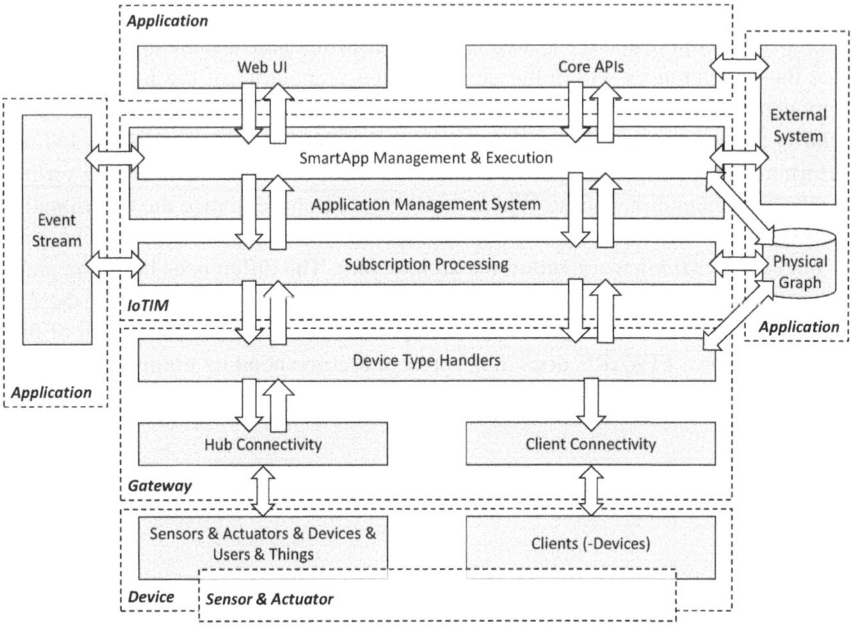

Fig. 9 SmartThings architecture based on [33]

Regarding SmartThings, our IoT reference architecture covers each component of the architecture. As described above, the *Sensor*, *Actuator*, and *Device* components are overlapping, since SmartThings uses them within composed components. Nevertheless, this is appropriate to our definition.

3.9 Summary of the Comparison

Reflecting each comparison described above, our introduced IoT reference architecture is represented within each considered platform. Table 1 shows a summarized overview of the comparison: The rows are defined by the components of our reference architecture and the columns display each considered IoT platform, whereby the table cells indicate each component of the IoT platform matching to our reference architecture components.

Only the architecture of OpenMTC and SmartThings represent a *Sensor* and *Actuator* component. All remaining platforms, besides the Microsoft Azure IoT Hub, just mention those components within their documentation. The *Device* component is not represented within the architecture of OpenMTC and the IBM Watson IoT Platform, but mentioned within the documentation. The remaining platforms represent a *Device* component within their architecture. Furthermore, the platforms FIWARE, AWS IoT, and the Microsoft Azure IoT Hub further distinguish the concept of "Intelligent" Devices, which have already some kind of logical functionality included. Following, those "Intelligent" Devices are covered by our *Device*, *Gateway*, and *IoT Integration Middleware* components, respective to the level of integrated logical functionality. Since those differences within the definition and granularity of the IoT platforms' components are present, we mapped them onto our clearly separated components, to clarify the concept and the used granularity. Besides SiteWhere and AWS IoT, all platforms represent the concept of our *Gateway* within their architectures. Nevertheless, the documentations of SiteWhere and AWS IoT also embrace the functionality of our *Gateway*. Obviously, each platform represents the core functionality, i.e., our *IoT Integration Middleware* within the architecture. The differences lie in the granularity and the number of the components comprising the functionality of the *IoT Integration Middleware*. Furthermore, each platform enables the connection of further *Applications*. FIWARE does not represent a corresponding component within its architecture, but it is mentioned within the documentation.

4 Related Work

In this section, the IoT reference architecture is related to previously published IoT architectures, architecture reference models, domain models, and taxonomies.

Bauer et al. [5] describe seven functional components between a device and an application layer as part of an IoT reference architecture. The components are

Table 1 IoT platform comparison summary

Platform	FIWARE	OpenMTC	SiteWhere	Webinos	AWS IoT	IBM Watson IoT Platform	MS Azure IoT Hub	Samsung SmartThings
Sensor / Actuator	-*	Sensors & Actuators	-*	-*	-*	-*	-	Sensors & Actuators & Devices & Users & Things
Device	Device / NGSI Device	-*	Data from Devices + Commands to Devices	PZP: Policy + Session + Discovery + Context Manager	Things	-*	Device	Sensors & Actuators & Devices & Users & Things + Clients (-Devices)
Gateway	IoT Edge	Front-End: Core Features – Connectivity + Back-End: Connectivity	-*	PZP: Sync + Messaging Manager + PZH: Sync Manager	-*	Connect	Cloud Protocol Gateway + Field Gateway	Hub and Client Connectivity + Device Type Handlers
IoT Integration Middleware	IoT Back-End + Data Context Broker	Back-End: Connectivity + Core Features + Application Enablement	SiteWhere Tenant Engine	PZH: User Authentication + Policy Repository + Policy Enforcement + Web APIs	Message Broker + Thing Shadows + Thing Registry + Rules Engine + Security & Identity	Analytics + Risk Management + Connect + Information Management + Bluemix Open Standards Based Services + Flexible Deployment	IoT Hub + Event Processing and Insight + Device Business Logic, Connectivity Monitoring + Application Device Provisioning and Management	Application Management System + Subscription Processing
Application	-*	Applications + Other M2M Platform	Integration- + REST-components	Third Party Applications	Amazon Services + IoT Applications	IoT Industry Solutions + Third Party Apps	Application Device Provisioning and Management	Event Stream + Web UI + Core APIs + External System + Physical Graph

* Not represented in the figure of the architecture, but described within the documentation.

the Management, Service Organization, IoT Process Management, Virtual Entity, IoT Service, Security, and Communication. The Communication component can be mapped onto the *Gateway* of the presented IoT reference architecture in this work, while the remaining components build the *IoT Integration Middleware*, respectively. In contrast to our work, the *Sensor*, *Actuator*, *Device*, and *Application* components are not specifically defined.

Fremantle [14] introduces an IoT reference architecture comprising of five layers. The device layer encompasses *Devices*, *Sensors*, and *Actuators*, but does not detail the latter two in particular. The relevant transports layer abstracts the same concept as our *Gateway*. The aggregation/bus layer as well as the event processing and analytics layer correspond with our *IoT Integration Middleware*. Thus, they provide the core functionality of an IoT platform. Finally, further *Applications* as presented in this work are subsumed by Fremantle as client and external communications. Since this IoT reference architecture lacks unambiguous definitions of all components it does not pursue our goal to provide a clear terminology to understand commonalities and differences of diverse IoT platforms, it is less effective than our reference architecture.

Cisco [10] introduces a seven-layered IoT reference model. The *Devices*, *Sensors*, and *Actuators* as presented in this work are comprised in the Physical Devices and Controllers of Cisco's reference model, while the *Gateway* layer equals their Connectivity concept. The Edge (Fog) Computing, Data Accumulation, and Data Abstraction layer corresponds to the *IoT Integration Middleware* of our IoT reference architecture, whereas the Application layer corresponds roughly to the combination of the components *IoT Integration Middleware* and *Application*. Finally, the capability to connect arbitrary *Applications* to the *IoT Integration Middleware* is reflected by the concepts Collaboration and Processes by Cisco. Since the concepts introduced by Cisco are not unambiguously defined in their reference model, the concepts presented in this work can be used to exactly determine the meaning of Cisco's concepts by mapping the reference model by Cisco onto our IoT reference architecture.

The three-layer architecture by Zheng et al. [37] contains similar concepts as those outlined in our reference architecture and is also basis for the works by Wu et al. [35], Atzori et al. [4], and Aazam et al. [1]. Gathering data from and acting on the physical world is described by the abstract concept of a Perception Layer and corresponds with the combination of our *Sensors*, *Actuators*, and *Devices*. Pre-processing of gathered data and transmission to an integrating middleware is covered by the Network Layer, which corresponds to the interplay of *Device* and *Gateway* in our IoT reference architecture. The Application Layer is also a more coarse-grained concept and reflects the core functionality of the platform. Thus, it maps onto our *IoT Integration Middleware* and *Applications*. Further approaches by Atzori et al. [3, 4] and Xu et al. [36] are similar layered architectures and grasp the field of IoT from a service-oriented architecture (SOA) perspective. While they focus on the design of IoT architectures, they lack a clear and unambiguous definition of the concepts, which they introduce and rely on. Neither of these works map their introduced concepts onto existing technologies and platforms, which is one contribution of our work.

Kim et al. [21] investigated diverse IoT applications and abstracted a generic platform model from them. They introduce the concept of Things, which are closely related to *Devices* as presented in this work. Gateways provide connections to a Platform in cases that Things cannot communicate directly with the Platform. Service Users as well as Service and Software Providers are connected to the Platform by RESTful APIs. In cases where no complex data processing is required on the Platform, a Service Use can also connect directly to devices, e.g., to gather metering data. All components of this model are covered of our reference architecture, besides the user.

The IoT Reference Model discussed by Krčo et al. [23] is based upon the IoT Domain Model by Haller et al. [18]. The concepts Augmented Entity, User, Device, Resource, and Service are introduced. A definition of these concepts is given but it is not abstract and unambiguous enough for mapping different IoT platforms upon each other to foster their understanding. For instance, on the one hand, a device is described as a hardware component, which integrates sensors and/or actuators and is, therefore, responsible for monitoring and interacting with real-world objects. On the other hand, a device is also capable of connecting to further IT systems. This example shows that the concept of a device is only roughly defined, thus, it is unclear if the device may also takes on the role of a gateway or if such an indirection is not foreseen, which implies that devices always communicate directly with the platform.

Mineraud et al. [26] review 39 existing IoT platforms according to six criteria including for instance data ownership or developer support. Concerning the architecture, they distinguish between cloud-based and local IoT platforms, but they do not provide a detailed analysis of the architectures as we do.

A high-level taxonomy for the components of an IoT platform is introduced by Gubbi et al. [16]. It contains the components hardware, middleware, and presentation. Hardware covers sensors, actuators, and embedded communication hardware, while middleware covers on-demand storage and computing tools for data analytics, and presentation provides visualization and interpretation tools. However, the taxonomy is very coarse-grained and not detailed enough to foster a clear understanding of the introduced concepts, which leads to possibly diverse interpretations.

5 Conclusion

The IoT is slowly turning from vision into reality: IoT platforms play a central role within this evolution by providing significant building blocks. A lack of standardization and development in silos has led to a heterogeneous platform landscape. We argue that, as a result of this heterogeneity, comparing and selecting one of these platforms is a difficult task. Not only do they use different concepts and technologies, but also the terminology is not clearly defined. Many concepts and parts of these platforms are described with synonyms or homonyms, or differ in granularity.

To help with these problems, we introduced an IoT reference architecture which is based on existing platforms. We defined each component and described the communication between them. These components do not necessarily have to stay separated, but can be combined. We compared our reference architecture to eight existing platforms, four of which are open-source. We showed that the components of our architecture map to those of the existing platforms. When comparing or evaluating these different platforms, our IoT reference architecture can be a useful tool. Besides, it may be useful by providing a common basis on which to base new IoT platform designs.

Future work could present a technical definition of the reference architecture. Furthermore, this work will build the basis for a decision support approach, which provides a selection of IoT platforms based on user-given preferences. This will help a user to determine a suitable IoT solution for his case.

Acknowledgements The research leading to these results has received funding from the German government through the BMWi projects SmartOrchestra (01MD16001F) and NEMAR (03ET4018).

References

1. Aazam, M., I. Khan, A.A. Alsaffar, and E.N. Huh. 2014. Cloud of things: Integrating internet of things and cloud computing and the issues involved. In *International Bhurban conference on applied sciences and technology*. IEEE.
2. Amazon Web Services: AWS IoT Documentation. 2016. https://aws.amazon.com/de/documentation/iot/.
3. Atzori, L., A. Iera, and G. Morabito. 2010. The Internet of Things: A survey. *Computer Networks* 54(15): 2787–2805.
4. Atzori, L., A. Iera, G. Morabito, and M. Nitti. 2012. The Social Internet of Things (SIoT)—When social networks meet the Internet of Things: Concept, architecture and network characterization. *Computer Networks* 56(16): 3594–3608.
5. Bauer, M., M. Boussard, N. Bui, J.C.M. De Loof, S. Meissner, A. Nettsträter, J. Stefa, M. Thoma, and J.W. Walewski. 2013. IoT reference architecture. In *Enabling things to talk: Designing IoT solutions with the IoT architectural reference model*. Berlin: Springer.
6. Betts, D. 2016. Microsoft Azure—Übersicht über Azure IoT Hub. https://azure.microsoft.com/de-de/documentation/articles/iot-hub-what-is-iot-hub/.
7. Bonomi, F., R. Milito, P. Natarajan, J. Zhu. 2014. Fog computing: A platform for internet of things and analytics. In *Big data and internet of things: A roadmap for smart environments*, 169–186. Springer.
8. Borgia, E. 2014. The Internet of Things vision: Key features, applications and open issues. *Computer Communications* 54: 1–31.
9. Boswarthick, D., O. Elloumi, and O. Hersent, (eds.). 2012. M2M Communications. Wiley.
10. Cisco: The Internet of Things Reference Model. 2014. http://cdn.iotwf.com/resources/71/IoT_Reference_Model_White_Paper_June_4_2014.pdf.
11. Farhangi, H. 2010. The path of the smart grid. *IEEE Power and Energy Magazine* 8(1): 18–28.
12. FIWARE: FIWARE Wiki. 2016. https://forge.fiware.org/plugins/mediawiki/wiki/fiware/index.php/Main_Page.
13. Fraunhofer FOKUS: OpenMTC Platform Architecture. 2016. http://www.open-mtc.org/index.html#MainFeatures.
14. Fremantle, P. 2015. A reference architecture for the Internet of Things. http://wso2.com/wso2_resources/wso2_whitepaper_a-reference-architecture-for-the-internet-of-things.pdf.
15. Fuhrhop, C., J. Lyle, and S. Faily. 2012. The webinos project. In *Proceedings of the 21st international conference on World Wide Web*, 259–262. ACM.
16. Gubbi, J., R. Buyya, and S. Marusic. 2013. Internet of Things (IoT): A vision, architectural elements, and future directions. *Future Generation Computer Systems* 29(7): 1645–1660.

17. Guth, J., U. Breitenbücher, M. Falkenthal, F. Leymann, and L. Reinfurt. 2016. Comparison of IoT platform architectures: A field study based on a reference architecture. In *Proceedings of the international conference on cloudification of the Internet of Things (CIoT)* IEEE.
18. Haller, S.A.S., M. Bauer, and F. Carrez. 2013. A domain model for the Internet of Things. In *Proceedings of the IEEE international conference on green computing and communications and IEEE Internet of Things and IEEE Cyber, physical and social computing*. IEEE.
19. IBM: IBM Internet of Things Architecture Overview. 2016. https://www.iot-academy.info/mod/page/view.php?id=478.
20. Kargupta, H. 2012. Connected cars: How distributed data mining is changing the next generation of vehicle telematics products. In *International conference on sensor systems and software*. Springer.
21. Kim, J., J. Lee, J. Kim, and J. Yun. 2014. M2M service platforms: Survey, issues, and enabling technologies. *IEEE Communications Surveys & Tutorials* 16(1): 61–76.
22. Kopp, O., M. Falkenthal, N. Hartmann, F. Leymann, H. Schwarz, and J. Thomsen. 2015. Towards a cloud-based platform architecture for a decentralized market agent. In *Informatik 2015*, Lecture Notes in Informatics (LNI). Gesellschaft für Informatik e.V. (GI).
23. Krčo, S., B. Pokrić, and F. Carrez. 2014. Designing IoT and architecture(s). In *Proceedings of the IEEE World Forum on Internet of Things (WF-IoT)*. IEEE.
24. Lucke, D., C. Constantinescu, and E. Westkämper. 2008. Smart factory–A step towards the next generation of manufacturing. In *Manufacturing systems and technologies for the new frontier*, 115–118. Springer.
25. Merriam-Webster: Full definition of SENSOR. 2016. http://www.merriam-webster.com/dictionary/sensor.
26. Mineraud, J., O. Mazhelis, X. Su, and S. Tarkoma. 2016. A gap analysis of Internet-of-Things platforms. *Computer Communications* 89–90: 5–16.
27. Miorandi, D., S. Sicari, F. De Pellegrini, and I. Chlamtac. 2012. Internet of things: Vision, applications and research challenges. *Ad Hoc Networks* 10(7): 1497–1516.
28. Open Mobile Alliance Ltd.: NGSI Context Management. 2012. http://technical.openmobilealliance.org/Technical/release_program/docs/NGSI/V1_0-20120529-A/OMA-TS-NGSI_Context_Management-V1_0-20120529-A.pdf.
29. Postscapes: IoT Cloud Platform Landscape. Vendor List. 2016. http://www.postscapes.com/internet-of-things-platforms/.
30. Ricquebourg, V., D. Menga, D. Durand, B. Marhic, L. Delahoche, and C. Loge. 2006. The smart home concept: Our immediate future. In *1st IEEE international conference on e-learning in industrial electronics*. IEEE.
31. Schaffers, H., N. Komninos, M. Pallot, B. Trousse, M. Nilsson, and A. Oliveira. Smart cities and the future internet: Towards cooperation frameworks for open innovation. In *The future internet assembly*. Springer.
32. SiteWhere LLC.: SiteWhere System Architecture. 2016. http://documentation.sitewhere.org/architecture.html.
33. SmartThings, Inc.: SmartThings Documentation. 2016. http://docs.smartthings.com/en/latest/architecture/index.html.
34. Webinos Project: webinos project deliverable—Phase II architecture and components. 2012. Technical report.
35. Wu, M., T.J. Lu, F.Y. Ling, J. Sun, and H.Y. Du. 2010. Research on the architecture of Internet of Things. In *Proceedings of the 3rd International Conference on Advanced Computer Theory and Engineering (ICACTE)*. IEEE.
36. Xu, L.D., W. He, and S. Li. 2014. Internet of things in industries: A survey. *IEEE Transactions on Industrial Informatics* 10(4).
37. Zheng, L., H. Zhang, W. Han, X. Zhou, J. He, Z. Zhang, Y. Gu, and J. Wang. 2009. Technologies, applications, and governance in the Internet and of Things. In *Internet of Things—Global Technological and Societal Trends*. River Publishers.
38. Zorzi, M., A. Gluhak, S. Lange, and A. Bassi. 2010. From today's INTRAnet of things to a future INTERnet of things: A wireless- and mobility-related view. *IEEE Wireless Communications* 17(6): 44–51.

Fog Computing: A Taxonomy, Survey and Future Directions

Redowan Mahmud, Ramamohanarao Kotagiri and Rajkumar Buyya

Abstract In recent years, the number of Internet of Things (IoT) devices/sensors has increased to a great extent. To support the computational demand of real-time latency-sensitive applications of largely geo-distributed IoT devices/sensors, a new computing paradigm named "Fog computing" has been introduced. Generally, Fog computing resides closer to the IoT devices/sensors and extends the Cloud-based computing, storage and networking facilities. In this chapter, we comprehensively analyse the challenges in Fogs acting as an intermediate layer between IoT devices/sensors and Cloud datacentres and review the current developments in this field. We present a taxonomy of Fog computing according to the identified challenges and its key features. We also map the existing works to the taxonomy in order to identify current research gaps in the area of Fog computing. Moreover, based on the observations, we propose future directions for research.

1 Introduction

Fog computing is a distributed computing paradigm that acts as an intermediate layer in between Cloud datacentres and IoT devices/sensors. It offers compute, networking and storage facilities so that Cloud-based services can be extended closer to the IoT devices/sensors [1]. The concept of Fog computing was first introduced by Cisco in 2012 to address the challenges of IoT applications in conventional Cloud computing. IoT devices/sensors are highly distributed at the edge of the network along with real-time and latency-sensitive service requirements. Since Cloud datacentres are geographically centralized, they often fail to deal with storage and processing

R. Mahmud (✉) · R. Kotagiri · R. Buyya
Cloud Computing and Distributed Systems (CLOUDS) Laboratory Department of Computing and Information System, The University of Melbourne, Parkville, VIC 3010, Australia
e-mail: mahmudm@student.unimelb.edu.au

R. Kotagiri
e-mail: kotagiri@unimelb.edu.au

R. Buyya
e-mail: rbuyya@unimelb.edu.au

B. Di Martino et al. (eds.), *Internet of Everything*, Internet of Things,
https://doi.org/10.1007/978-981-10-5861-5_5

Fig. 1 Fog computing environment

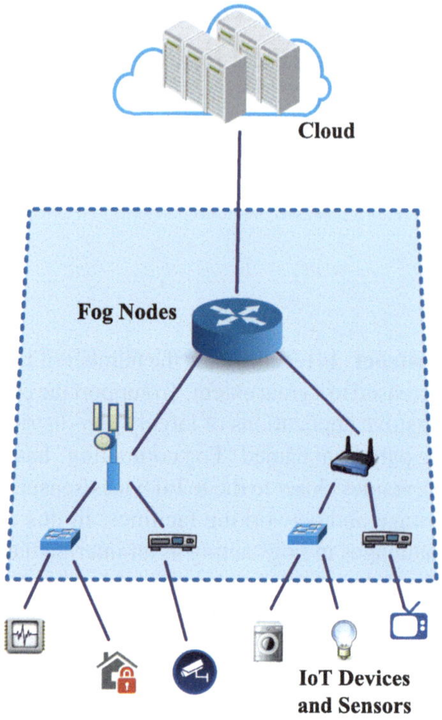

demands of billions of geo-distributed IoT devices/sensors. As a result, congested network, high latency in service delivery, poor Quality of Service (QoS) are experienced [2].

Typically, a Fog computing environment is composed of traditional networking components e.g. routers, switches, set top boxes, proxy servers, Base Stations (BS), etc. and can be placed at the closer proximity of IoT devices/sensors as shown in Fig. 1. These components are provided with diverse computing, storage, networking, etc. capabilities and can support service-applications execution. Consequently, the networking components enable Fog computing to create large geographical distributions of Cloud-based services. Besides, Fog computing facilitates location awareness, mobility support, real-time interactions, scalability and interoperability [3]. Thereby, Fog computing can perform efficiently in terms of service latency, power consumption, network traffic, capital and operational expenses, content distribution, etc. In this sense, Fog computing better meets the requirements with respect to IoT applications compared to a solely use of Cloud computing [4]. However, the concept of Fog computing is very much similar to the existing computing paradigms. In this chapter, we elaborately discuss the fundamental differences of Fog computing with other computing paradigms. Here, we also analyse different aspects of Fog computing including corresponding resource architecture, service quality, security issues, etc. and review recent research works from the literature. We present a taxonomy based on the key properties and associated challenges in Fog computing. We map the existing works to the taxonomy to identify innovative approaches and limitations in this field. Based on the observations, we also propose potential future directions so that further improvement in Fog computing can be achieved. The rest of the chapter is organized as follows. In Sect. 2, we discuss the differences of Fog computing with other related computing approaches. After that, we describe the challenges of Fog computing and propose our taxonomy in Sects. 3 and 4, respectively. From Sects. 5 to 10, we map the existing research works to the proposed taxonomy. In Sect. 11, we analyze research gaps and present some promising directions towards future research in Fog computing. Finally, we summarize the findings and conclude the paper.

2 Related Computing Paradigms

With the origination of Cloud computing, computation technology has entered to a new era. Many computation service providers including Google, Amazon, IBM, Microsoft, etc. are currently nurturing this popular computing paradigm as a utility. They have enabled cloud based services such as Infrastructure as a Service (IaaS), Platform as a Service (PaaS), Software as a Service (SaaS), etc. to handle numerous enterprise and educational related issues simultaneously. However, most of the Cloud datacentres are geographically centralized and situated far from the proximity of the end devices/users. As a consequence, real-time and latency-sensitive computation service requests to be responded by the distant Cloud datacentres often endure large round-trip delay, network congestion, service quality degradation, etc. To resolve these issues besides centralized Cloud computing, a new concept named "Edge computing" has recently been proposed [5].

The fundamental idea of Edge computing is to bring the computation facilities closer to the source of the data. More precisely, Edge computing enables data processing at the edge network [6]. Edge network basically consists of end devices (e.g. mobile phone, smart objects, etc.), edge devices (e.g. border routers, set-top boxes, bridges, base stations, wireless access points etc.), edge servers, etc. and these components can be equipped with necessary capabilities for supporting edge computation. As a localized computing paradigm, Edge computing provides faster responses to the computational service requests and most often resists bulk raw data to be sent towards core network. However, in general Edge computing does not associate IaaS, PaaS, SaaS and other cloud based services spontaneously and concentrate more towards the end devices side [7].

Taking the notion of Edge and Cloud computing into account, several computing paradigms have already been introduced in computation technology. Among them Mobile Edge Computing (MEC), Mobile Cloud Computing (MCC) are considered as the potential extensions of Cloud and Edge computing.

As an edge-centric computing paradigm, MEC has already created significant buzz in the research domain. MEC has been regarded as one of the key enablers of modern evolution of cellular base stations. It offers edge servers and cellular base staions to be operated combinedly [8]. MEC can be either connected or not connected with distant Cloud datacentres. Thus along with end mobile devices, MEC supports 2 or 3 tire hierarchical application deployment in the network [9]. Besides, MEC targets adaptive and faster initiation of cellular services for the customers and enhances network efficiency. In recent times, significant improvement in MEC has been made so that it can support 5G communications. Moreover, it aims at flexible access to the radio network information for content distribution and application development [10, 11].

MCC is another recent trend in computation. Due to the proliferation of smart mobile devices, nowadays end users prefer to run necessary applications in their hand held devices rather than traditional computers. However, most of the smart mobile devices are subjected to energy, storage and computational resource constraints [12]. Therefore, in critical scenarios, it is more feasible to execute compute intensive applications outside the mobile devices compared to execute those applications locally. In this case, MCC provides necessary computational resources to support such remote execution of offloaded mobile applications in closer proximity of end users [13, 14]. In MCC, most often light-weight cloud servers named cloudlet [15] are placed at the edge network. Along with end mobile devices and Cloud datacentres, cloudlets develop a 3 tire hierarchical application deployment platform for rich mobile applications [9]. In brief, MCC combines cloud computing, mobile computing and wireless communication to enhance Quality of Experience (QoE) of mobile users and creates new business opportunities for both network operators and cloud service providers.

Like MEC and MCC, Fog computing can also enable edge computation. However, besides edge network, Fog computing can be expanded to the core network as well [3]. More precisely, both edge and core networking (e.g. core routers, regional servers, wan switches, etc.) components can be used as computational infrastructure in Fog computing (Fig. 2). As a consequence, multi-tire application deployment and service demand mitigation of huge number of IoT devices/sensors can easily be observed through Fog computing. In addition, Fog computing components at the edge network can be placed closer to the IoT devices/sensors compared to cloudlets and cellular edge servers. As IoT devices/sensors are densely distributed and require real-time responses to the service requests, this approach enables IoT data to be stored and processed within the vicinity of IoT device/sensors. As a result, service delivery latency for real-time IoT applications will be minimized to a great extent. Unlike Edge computing, Fog computing can extend cloud based services like IaaS, PaaS, SaaS, etc. to the edge of the network as well. Due to the aforementioned features, Fog computing is considered as more potential and well structured for IoT compared to other related computing paradigms.

3 Challenges in Fog Computing

Fog computing is considered as the promising extension of Cloud computing paradigm to handle IoT related issues at the edge of network. However, in Fog computing, computational nodes are heterogeneous and distributed. Besides, Fog based services

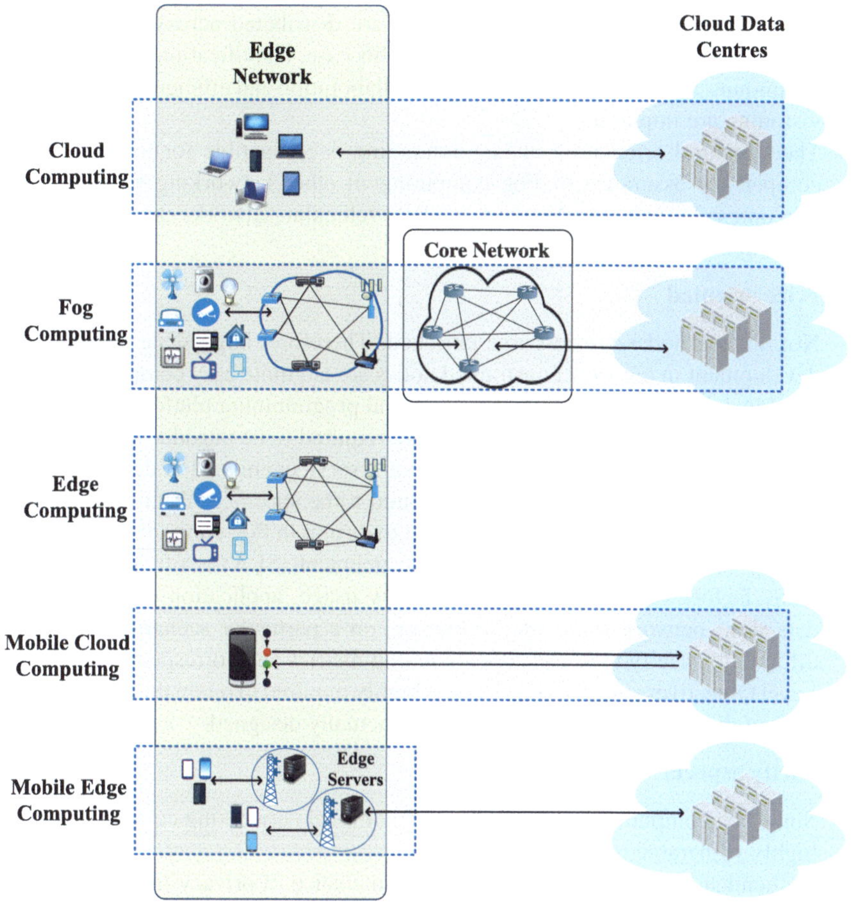

Fig. 2 Computation domain of Cloud, Fog, Edge, Mobile Cloud and Mobile Edge computing

have to deal with different aspects of constrained environment. Assurance of security is also predominant in Fog computing. Analysing the features of Fog computing from structural, service oriented and security perspectives, the challenges in this field can be listed as follows:

- **Structural issues**
  - Different components from both edge and core network can be used as potential Fog computing infrastructure. Typically these components are equipped with various kinds of processors but are not employed for general purpose computing. Provisioning the components with general purpose computation besides their traditional activities will be very challenging.
  - Based on operational requirements and execution environment, the selection of suitable nodes, corresponding resource configuration and places of deployment are vital in Fog as well.

- In Fog computing, computational nodes are distributed across the edge network and can be virtualized or shared. In this case, identification of appropriate techniques, metrics, etc. for inter-nodal collaboration and efficient resource provisioning are important.
- The structural orientation of Fog computing is compatible for IoT. However, competency assurance of Fog computing in other networking systems such as Content Distribution Network (CDN), vehicular network, etc. will be very challenging.

• **Service oriented**

- Not all Fog nodes are resource enriched. Therefore, large scale applications development in resource constrained nodes are not quite easy compared to conventional datacentres. In this case, potential programming platform for distributed applications development in Fog are required to be introduced.
- Policies to distribute computational tasks and services among IoT devices/sensors, Fog and Cloud infrastructures are required to be specified. Data visualization through web-interfaces are also difficult to design in Fog computing.
- In Fog computing, the Service Level Agreement (SLA) is often affected by many factors such as, service cost, energy usage, application characteristics, data flow, network status etc. Therefore, on a particular scenario, it is quite difficult to specify the service provisioning metrics and corresponding Service Level Objectives (SLOs). Besides, it is highly required to retain the fundamental QoS of the Fog nodes for which they are actually designed.

• **Security aspects**

- Since Fog computing is designed upon traditional networking components, it is highly vulnerable to security attacks.
- Authenticated access to services and maintenance of privacy in a largely distributed paradigm like Fog computing are hard to ensure.
- Implementation of security mechanisms for data-centric integrity can affect the QoS of Fog computing to a great extent.

In addition to aforementioned challenges service scalability, end users QoE, context-awareness, mobility support are very crucial performance indicator for Fog computing and very difficult to deal with in real-time interactions.

4 Taxonomy

Figure 3 represents our proposed taxonomy for Fog computing. Based on the identified challenges from Sect. 3, the taxonomy provides a classification of the existing works in Fog computing. More precisely, the taxonomy highlights the following aspects in Fog computing.

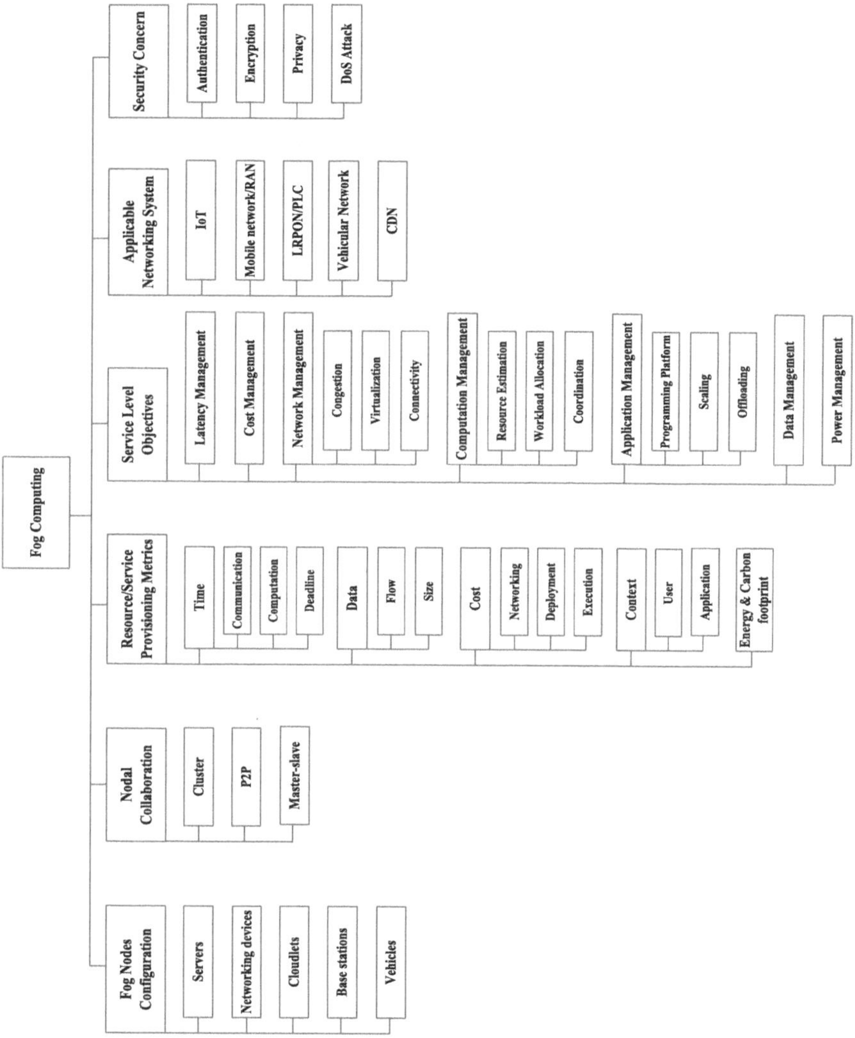

Fig. 3 Taxonomy of Fog Computing

- Fog Nodes Configuration. The computational nodes with heterogeneous architecture and configurations that are capable to provide infrastructure for Fog computing at the edge of the network.
- Nodal Collaboration. The techniques for managing nodal collaboration among different Fog nodes within the edge network.
- Resource/Service Provisioning Metric. The factors that contribute to provision resources and services efficiently under different constraints.
- Service Level Objectives. The SLOs that have been attained by deploying Fog computing as an intermediate layer between Cloud datacentres and end devices/sensors.
- Applicable Network System. The different networking systems where Fog computing has been introduced as extension of other computing paradigms.
- Security Concern. The security issues that have been considered in Fog computing on different circumstances.

Proposed system and corresponding solutions in the existing works generally covers different categories of the taxonomy. However, as this taxonomy is designed based on the associated features of Fog computing, it does not reflect the relative performance of the proposed solutions. Actually, the reviewed existing works considers diverse execution environment, networking topology, application characteristics, resource architecture, etc. and targets different challenges. Therefore, identification of the best approach for Fog computing in terms of structural, service and security aspects is very difficult.

In the following sections (from Sects. 5 to 10), we map the existing works in Fog computing to our proposed taxonomy and discuss different facts in detail.

5 Fog Nodes Configuration

Five types of Fog nodes and their configurations have been mentioned in the literature: namely servers, networking devices, cloudlets, base stations, vehicles.

5.1 Servers

The Fog servers are geo-distributed and are deployed at very common places for example; bus terminals, shopping centres, roads, parks, etc. Like light-weight Cloud servers, these Fog servers are virtualized and equipped with storage, compute and networking facilities. There are many works that have considered Fog servers as main functional component of Fog computing.

In some papers based on the physical size, Fog servers are termed as micro servers, micro datacentres [16, 17], nano servers [18], etc. whereas other works categorize Fog servers based on their functionalities like cache servers [19], computation servers, storage servers [20], etc. Server based Fog node architecture enhances the computation and storage capacity in Fog computing. However, it limits the pervasiveness of the execution environment.

5.2 Networking Devices

Devices like gateway routers, switches, set-top boxes, etc. besides their traditional networking activities (routing, packet forwarding, analog to digital signal conversions, etc.) can act as potential infrastructure for Fog computing. In some existing works, the networking devices are designed with certain system resources including data processors, extensible primary and secondary memory, programming platforms, etc. [21, 22].

In other works, apart from conventional networking devices, several dedicated networking devices like *Smart gateways* [23], *IoT Hub* [24] have been introduced as Fog nodes. Distributed deployment of networking devices helps Fog computing to be ubiquitous although physical diversity of the devices significantly affects service and resource provisioning.

5.3 Cloudlets

Cloudlets are considered as micro-cloud and located at the middle layer of end device, cloudlet, and Cloud hierarchy. Basically cloudlets have been designed for extending Cloud based services towards mobile device users and can complement MCC [12]. In several works [25, 26], cloudlets are mentioned as Fog nodes. Cloudlet based Fog computing are highly virtualized and can deal with large number of end devices simultaneously. In some cases, due to structural constraints, cloudlets even after deploying at the edge act as centralized components. In this sense, the limitations of centralized computation still remain significant in Fog computing which resist to support IoT.

5.4 Base Stations

Base stations are very important components in mobile and wireless networks for seamless communication and data signal processing. In recent works, traditional base stations equipped with certain storing and computing capabilities are considered suitable for Fog computing [27, 28]. Like traditional base stations, Road Side Unit (RSU) [29] and small cell access points [30], etc. can also be used as potential Fog nodes.

Base stations are preferable for Fog based extension of Cloud Radio Access Network (CRAN), Vehicular Adhoc Network (VANET), etc. However, formation of a dense Fog environment with base stations is subjected to networking interference and high deployment cost.

5.5 Vehicles

Moving or parked vehicles at the edge of network with computation facilities can serve as Fog nodes [31, 32]. Vehicles as Fog nodes can form a distributed and highly scalable Fog environment. However, the assurance of privacy and fault tolerance along with desired QoS maintenance will be very challenging in such environment.

6 Nodal Collaboration

Three techniques (cluster, peer to peer and master-slave) for nodal collaboration in Fog computing have been specified in the literature.

6.1 Cluster

Fog nodes residing at the edge can maintain a collaborative execution environment by forming cluster among themselves. Clusters can be formed either based on the homogeneity of the Fog nodes [26] or their location [31]. Computational load balancing [33] and functional sub-system development [34] can also be given higher priority while forming cluster among the nodes.

Cluster based collaboration is effective in exploiting capabilities of several Fog nodes simultaneously. However, static clusters are difficult to make scalable in runtime and dynamic formation of clusters largely depends on the existing load and the availability of Fog nodes. In both cases networking overhead plays a vital role.

6.2 Peer to Peer

In Fog computing Peer to Peer (P2P) collaboration among the nodes is very common. P2P collaboration can be conducted in both hierarchical [21] and flat order [35]. Besides based on proximity, P2P collaboration between Fog nodes can be classified as home, local, non-local, etc. [18]. Through P2P collaboration not only processed output of one node appears as input to another node [36] but also virtual computing instances are shared between the nodes [25].

Augmentation of Fog nodes in P2P collaboration is quite simple and nodes can be made reusable. However, reliability and access control related issues are predominant in P2P nodal collaboration.

6.3 *Master-Slave*

In several works, master-slave based nodal collaboration has been mentioned elaborately. Through master-slave based collaboration generally a master Fog node controls functionalities, processing load, resource management, data flow, etc. of underlying slave nodes [16].

Besides, master-slave based approach along with cluster and P2P based nodal interactions can form a hybrid collaborative network within the Fog computing environment [22, 29]. However, in real-time data processing due to this kind of functional decomposition, the master and the slave Fog nodes require high bandwidth to communicate with each other.

7 Resource/Service Provisioning Metrics

In existing literature of Fog computing many factors including time, energy, user-application context, etc. have been found playing important roles in resource and service provisioning.

7.1 *Time*

In Fog computing paradigm, time is considered as one of the important factors for efficient resource and service provisioning.

Computation time refers to the required time for task execution. Computation time of an application largely depends on resource configuration where the application is running or the task has been scheduled [20] and can be changed according to the existing load. Besides, task computation time helps to identify the active and idle periods of different applications which significantly influences resource and power management in Fog [18].

Communication time basically defines the networking delay to exchange data elements in Fog computing environment. In the literature it has been discussed in 2 folds: End device/sensors to Fog nodes [37], Fog nodes to Fog nodes [20]. Required communication time reflects the network context which assists selection of suitable Fog nodes for executing tasks.

Deadline specifies the maximum delay in service delivery that a system can tolerate. In some papers deadline satisfied task completion has been considered as important parameter for measuring QoS of the system [30, 32]. Basically service delivery deadline plays a significant role in characterizing latency sensitive and latency tolerant applications.

In addition, the impact of other time based metrics like data sensing frequency of end device/sensors, service access time in multi-tenant architecture, expected service response time, etc. can be investigated for efficient service and resource provisioning in Fog computing.

7.2 Data

Among data-centric metrics, input data size and data flow characteristics are found very common in Fog computing literature.

Data size points to the amount of data that has to be processed through Fog computing. In several works, data size has been discussed in respect of computational space requirements of the requests [35]. Besides, bulk data collected from distributed devices/sensors can contain the features of Big Data [22] as well. In this case, provisioning resource and service according to the data load can be an effective approach. Moreover, data size plays an important role in making decision about either local or remote processing of the corresponding computational tasks [38].

Data flow defines the characteristics of data transmission. Data flow through out the Fog computing environment can be event driven [16] or real time [36] and can influence resource and associated service provisioning to a great extent. Besides, sudden change in data flow sometimes promotes dynamic load balancing among the nodes [33].

Moreover, the effectiveness of heterogeneous data architecture, data semantic rules, data integrity requirements can also be studied for resource and service provisioning in Fog computing.

7.3 Cost

In certain cases, cost related factors form both service providers and users perspective become very influential in Fog resource and service provisioning.

Networking cost in Fog computing environment is directly related to the bandwidth requirements and associated expenses. In several works data uploading cost from end devices/sensors and inter-nodal data sharing cost have been considered as the elements of networking cost [28] whereas in other works, experienced network latency due to bandwidth issues has been termed as networking cost [39].

Deployment cost is basically associated with the infrastructure placement related expenses in Fog computing environment. In some papers cost-effective infrastructure deployment has been considered supportive for efficient resource and service provisioning. Infrastructure deployment cost can be discussed in terms of both placing

Fog nodes in the network [40] and creating virtual computing instances in Fog nodes [28].

Execution cost refers to the computational expenses of Fog nodes while running applications or processing tasks. Although in other computing paradigms execution cost is widely used in resource provisioning and billing, in Fog computing this metric has been used in very few works. In these works total execution cost has been calculated in terms of task completion time and per unit time resource usage cost [39].

In addition to aforementioned costs, expenses related to security safeguards, the maximum price that an user is willing to pay for a service, migration costs can also be considered for resource and service provisioning in Fog computing.

7.4 Energy Consumption and Carbon Footprint

In several works, energy related issues have been given higher priority in provisioning Fog resources and services. The energy consumption of all devices in home-based Fog computing environment [34] and the energy-latency tradeoff in different phases of Fog-Cloud interaction [41] have been highlighted widely in these works. In another work, carbon emission rate of different nodes in respect of per unit energy usage have been considered for resource provisioning purposes [42].

As end devices/sensors are energy-constrained, energy aspects of end components for example residual battery lifetime, energy-characteristics of communication medium can also be investigated in provisioning Fog resources.

7.5 Context

Context refers to situation or condition of a certain entity in different circumstances. In Fog based research works user and application level context have been discussed for resource and service provisioning.

User context such as user characteristics, service usage history, service relinquish probability, etc. can be used for allocating resources for that user in future [43]. Users service feedback for example Net Promoter Score (NPS) and user requirements [44] can also be used for service and resource provisioning purposes [45]. In other works users density [27], mobility [31] and network status [19] have also been considered for service provisioning.

Application context can be considered as operational requirements of different applications. Operational requirements includes task processing requirements (CPU speed, storage, memory) [23, 29, 46], networking requirements [24, 25], etc. and can affect resource and service provisioning. In other works current task load of different applications [26, 35] have also been considered as application context.

Moreover, contextual information in Fog computing can be discussed in terms of execution environment, nodal characteristics, application architecture, etc. and along with the other contexts they can play vital roles in provisioning resource and services. Therefore, it is essential to investigate the impact of every contextual information in-detail.

8 Service Level Objectives

In existing literature, several unique Fog node architecture, application programming platform, mathematical model and optimization technique have been proposed to attain certain SLOs. Most of the attained SLOs are management oriented and cover latency, power, cost, resource, data, application, etc. related issues.

8.1 Latency Management

Latency management in Fog computing basically resists the ultimate service delivery time from surpassing an accepted threshold. This threshold can be the maximum tolerable latency of a service request or applications QoS requirement.

To ensure proper latency management, in some works efficient initiation of nodal collaboration has been emphasized so that the computation tasks through the collaborated nodes can be executed within the imposed latency constraints [30]. In another work, computation task distribution between the client and Fog nodes have been conducted with a view to minimizing the total computation and communication latency of service requests [20].

Besides, in another work architecture of low-latency Fog network has been proposed for latency management [37]. The basic intention of this work is to select that node from the Fog network which provides lowest latency in service delivery.

8.2 Cost Management

Cost management in Fog computing can be discussed in terms of Capital Expenses (CAPEX) and Operating Expenses (OPEX).

The main contributor of CAPEX in Fog computing is the deployment of cost of distributed Fog nodes and their associated networks. In this case, suitable placement and optimized number of Fog nodes play a significant role in minimizing the CAPEX in Fog computing. Investigating this issue, a Fog computing network architecture has been proposed in [40] that minimizes the total CAPEX in fog computing by optimizing the places and numbers of Fog node deployment.

In another work [28], Fog nodes have been considered as virtualized platforms for launching VMs. Execution of data processing operations in these VMs are not free of cost and the cost can be varied from provider to provider. Therefore, it is feasible to exploit cost-diversity of different Fog nodes/ providers for minimizing OPEX in Fog computing. In respect to this fact, a solution to find suitable set of Fog nodes for hosting VMs has been proposed in that paper which aims to minimize the OPEX in Fog computing.

8.3 Network Management

Network management in Fog computing includes core-network congestion control, support for Software Define Network (SDN)/ Network Function Virtualization (NFV), assurance of seamless connectivity, etc.

Network congestion mainly occurs due to increasing overhead on the network. As in IoT, end devices/sensors are highly distributed across the edge, simultaneous interactions of end components with Cloud datacentres can increase the overhead on the core network to a great extent. In such case network congestion will occur and degrade the performance of the system. Taking cognizance of this fact, in [23] a layered architecture of Fog node has been proposed that provides local processing of the service requests. As a consequence, despite of receiving bulk service requests, Clouds get consized version of the requests which contribute less to the network congestion.

Virtualization of conventional networking system has already drawn significant research attention. SDN is considered as one of the key enablers of virtualized network. SDN is a networking technique that decouples the control plane from networking equipment and implements in software on separate servers. One of the important aspects of SDN is to provide support for NFV. Basically, NFV is an architectural concept that virtualizes traditional networking functions (network address translation (NAT), firewalling, intrusion detection, domain name service (DNS), caching, etc.) so that they can be executed through software. In Cloud based environment SDN and NFV is quite influencing due to their wide range of services. Being motivated by this, in several research works [16, 29, 38] new network structures of Fog computing have been proposed to enable SDN and NFV.

Connectivity ensures seamless communication of end devices with other entities like Cloud, Fog, Desktop computers, Mobile devices, end devices, etc. despite of their physical diversity. As a consequence, resource discovery, maintenance of communication and computation capacity become easier within the network. Several works in Fog computing have already targeted this issue and proposed new architecture of Fog nodes e.g. *IoT Hub* [24] and Fog networking e.g. *Vehicular Fog Computing* [31] for connectivity management and resource discovery. Besides, for secured connectivity among the devices a policy driven framework has also been developed for Fog computing [25].

8.4 Computation Management

Among the attained SLOs, assurance of proper computational resource management in Fog computing is very influential. Fog computing resource management includes resource estimation, workload allocation, resource coordination, etc.

Resource estimation in Fog computing helps to allocate computational resources according to some policies so that appropriate resources for further computation can be allocated, desired QoS can be achieved and accurate service price can be imposed. In existing literature, resource estimation policies are developed in terms of user characteristics , experienced QoE, features of service accessing devices, etc. [17, 43, 45].

Workload allocation in Fog computing should be done in such a way so that utilization rate of resources become maximized and longer computational idle period get minimized. More precisely, balanced load on different components is ensured. In a Fog based research work [20], scheduling based workload allocation policy has been introduced to balance computation load on Fog nodes and client devices. As a consequence overhead on both parts become affordable and enhance QoE. In another work [41] a workload allocation framework has been proposed that balances delay and power consumption in Fog-Cloud interaction.

Coordination among different Fog resources is very essential as they are heterogeneous and resource constrained. Due to decentralized nature of Fog computing, in most cases large scale applications are distributively deployed in different Fog nodes. In such scenarios without proper co-ordination of Fog resources, attainment of desired performance will not be very easy. Considering this fact, in [36] a directed graph based resource co-ordination model has been proposed for Fog resource management.

8.5 Application Management

In order to ensure proper application management in Fog computing, efficient programming platforms are very essential. Besides the scalability and computation offloading facilities also contribute significantly in application management.

Programming platform provides necessary components such as interfaces, libraries, run-time environment, etc. to develop, compile and execute applications. Due to dynamic nature of Fog computing, assurance of proper programming support for large-scale applications is very challenging. In order to overcome this issue, a new programming platform named *Mobile Fog* [21] has been introduced. *Mobile Fog* offers simplified abstraction of programming models for developing large-scale application over heterogeneous-distributed devices. In another paper [36], besides coordinating resources during applications execution, a programming platform based on distributed data flow approach has also been designed for application development in Fog computing.

Scaling points to the adaptation capability of applications in retaining their service quality even after proliferation of application users and unpredictable events. Scaling techniques can also be applied in application scheduling and users service access. To support scalable scheduling of data stream applications, architecture of a QoS-aware self adaptive scheduler [26] has been recently proposed in Fog computing. This scheduler can scale applications with the increasing of both users and resources and does not ask for global information about the environment. Moreover, due to self-adaptive capability of the scheduler, automatic reconfiguration of the resources and placement of applications in a distributed fashion become easier. Besides, based on distance, location and QoS requirements of the service accessing entities, an adaptive technique for users service access mode selection has also been proposed in Fog computing [27].

Offloading techniques facilitate resource constrained end devices in sending their computational tasks to some resource-enriched devices for execution. Computational offloading is very common in mobile cloud environment. However, recently, as a part of compatibility enhancement of Fog computing for other networking systems, computation offloading support for mobile applications in Fog computing have been emphasized in several papers. In these papers offloading techniques have been discussed in terms of both distributed computation of mobile applications [35] and resources availability [39].

8.6 Data Management

Data management is another important SLO that is highly required to be achieved for efficient performance of Fog computing. In different research works data management in Fog computing has been discussed from different perspectives. In [23, 44] initiation of proper data analytic services and resource allocation for data pre-processing have been focused for data management policy in Fog computing. Besides, low-latency aggregation of data coming from distributed end devices/sensors can also be considered for efficient data management [22]. Moreover, the storage capability of end devices/sensors are not so reach. In this case, storage augmentation in Fog computing for preserving data of end entities can be very influential. Therefore, in [39] besides application management, storage expansion in Fog computing for mobile devices have also been discussed as integral part of data management.

8.7 Power Management

Fog computing can be used as an effective platform for providing power management as a service for different networking systems. In [34], a service platform for Fog computing has been proposed that can enable power management in home based IoT network with customized consumer control. Additionally, Fog computing can manage power usage of centralized Cloud data centres in certain scenarios. Power consumed by Cloud datacentres largely depends on type of running applications. In this case, Fog computing can complement Cloud datacentres by providing infrastructure for hosting several energy-hungry applications. As a consequence energy consumption in Cloud datacentres will be minimized that eventually ensures proper power management for Cloud datacentres [18]. Moreover, by managing power in Fog computing emission of carbon footprint can also be controlled [42].

9 Applicable Network System

Fog computing plays a significant role in IoT. However, in recent research works the applicability of Fog computing in other networking systems (mobile network, content distribution network , radio access network, vehicular network, etc.) have also been highlighted.

9.1 Internet of Things

In IoT, every devices are interconnected and able to exchange data among themselves. IoT environment can be described from different perspectives. Besides specifying IoT as a network for device to device interaction [21, 24, 36], in several Fog based research works this interaction have been classified under industry [46] and home [34] based execution environment. Moreover, Wireless Sensors and Actuators Network [16], Cyber-Physical Systems [28], Embedded system network [20], etc. have also been considered as different forms of IoT while designing system and service models for Fog computing.

9.2 Mobile Network/Radio Access Network

Mobile network is another networking system where applicability of Fog computing has been explored through several research works. Basically, in these works much emphasize has been given on investigating the compatibility of Fog computing in 5G mobile networking [30, 33, 37]. 5G enables higher speed communication, signal capacity and much lower latency in service delivery compared to existing cellular systems. Besides 5G, Fog computing can also be applied in other mobile networks like 3G, 4G, etc. Moreover, in another work [41], power-delay tradeoff driven workload allocation in Fog-Cloud for mobile based communication has been investigated in detail. Radio Access Network (RAN) facilitates communication of individual devices with other entities of a network through radio connections. Cloud assisted RAN named CRAN has already drawn significant research attention. In order to complement CRAN, the potentiality of Fog computing based radio access network has also been explored in [38].

9.3 Long-Reach Passive Optical Network/Power Line Communication

Long-Reach Passive Optical Network (LRPON) has been introduced for supporting latency-sensitive and bandwidth-intensive home, industry, and wireless oriented backhaul services. Besides, covering a large area, LRPONs simplify network consolidation process. In [40], Fog computing has been integrated with LRPONs for optimized network design.

Power-line communication (PLC) is a widely used communication method in Smart Grid. In PLC, using electrical wiring both data and Alternating Current (AC) are simultaneously transmitted. Fog computing enabled PLC in electric power distribution has been discussed elaborately in [22].

9.4 Content Distribution Network

Content Distribution Network (CDN) is composed of distributed proxy servers that provide content to end-users ensuring high performance and availability. In several Fog based research works [19, 42], Fog nodes are considered as content servers to support content distribution through Fog computing. Since Fog nodes are placed in distributed manner across the edge of the network, Fog based content services can be accessed by the end users within a very minimal delay. As a consequence, high performance in content distribution will be easier to ensure.

9.5 Vehicular Network

Vehicular network enables autonomous creation of a wireless communication among vehicles for data exchange and resource augmentation. In this networking system vehicles are provided with computational and networking facilities. In several research works [29, 31, 32] vehicles residing at the edge network are considered as Fog nodes to promote Fog computing based vehicular network.

10 Security Concern

Security vulnerability of Fog computing is very high as it resides at the underlying network between end device/sensors and Cloud datacentres. However, in existing literature, security concerns in Fog computing has been discussed in terms of users authentication, privacy, secured data exchange, DoS attack, etc.

10.1 Authentication

Users authentication in Fog based services play an important role in resisting intrusion. Since Fog services are used in "pay as you go" basis, unwanted access to the services are not tolerable in any sense. Besides user authentication, in [25], device authentication, data migration authentication and instance authentication has also been observed for secured Fog computing environment.

10.2 Privacy

Fog computing processes data coming from end device/sensors. In some cases, these data are found very closely associated with users situation and interest. Therefore, proper privacy assurance is considered as one of the important security concerns in Fog computing. In [31] the challenges regarding privacy in Fog based vehicular computing have been pointed for further investigation.

10.3 Encryption

Basically, Fog computing complements Cloud computing. Data that has been processed in Fog computing, in some cases has to be forwarded towards Cloud. As these data often contains sensitive information, it is highly required to encrypt them in Fog nodes. Taking this fact into account, in [23], a data encryption layer has been included in the proposed Fog node architecture.

10.4 DoS Attack

Since, Fog nodes are resource constraint, it is very difficult for them to handle large number of concurrent requests. In this case, performance of Fog computing can be degraded to a great extent. To create such severe service disruptions in Fog computing,

Denial-of-Service (DoS) attacks can play vital roles. By making a lot of irrelevant service requests simultaneously, Fog nodes can be made busy for a longer period of time. As a result, resources for hosting useful services become unavailable. In [24], this kind of DoS attack in Fog computing has been discussed with clarification.

11 Gap Analysis and Future Directions

Fog computing resides at closer proximity of the end users and extends Cloud based facilities. In serving largely distributed end devices/sensors, Fog computing plays very crucial roles. Therefore, in recent years Fog computing has become one of the major fields of research from both academia and business perspectives. In Table 1, a brief summary of some reviewed papers from existing literature of Fog computing has been highlighted. Although many important aspects of Fog computing have been identified in the existing literature, there exist some other issues that are required to be addressed for further improvement of this field. In this section, the gaps from existing literatures along with several future research directions have been discussed.

11.1 Context-Aware Resource/Service Provisioning

Context-awareness can lead to efficient resource and service provisioning in Fog computing. Contextual information in Fog computing can be received in different forms, for example;

- Environmental context : Location, Time (Peak, Off-peak), etc.
- Application context : Latency sensitivity, Application architecture, etc.
- User context: Mobility, Social interactions, Activity, etc.
- Device context: Available resources, Remaining battery life time, etc.
- Network context: Bandwidth, Network traffic, etc.

Although several Fog based research works have considered contextual information in estimating resources, many important aspects of contextual information are still unexplored. Investigation of different techniques to apply contextual information in resource and service management can be a potential field towards Fog based research.

11.2 Sustainable and Reliable Fog Computing

Sustainability in Fog computing optimizes its economic and environmental influence to a great extent. However, the overall sustainable architecture of Fog computing is subject to many issues like assurance of QoS, service reusability, energy-efficient

Table 1 Review of state-of-art in Fog Computing

Work	Fog nodes	Nodal collaboration	Provisioning metrics	SLOs	Applicable network	Security concerns
Lee et al. [16]	Servers	Master-Slave	Data (flow)	Network management	IoT	–
Aazam et al. [17]	Servers	–	Context (user)	Resource management	IoT	–
Jalali et al. [18]	Servers	Peer to Peer	Time (computing)	Power management	CDN	–
Zhu et al. [19]	Servers	–	Context (user)	Application management	CDN	–
Zeng et al. [20]	Servers	Peer to Peer	Time (communication, computation)	Resource management	IoT	–
Hong et al. [21]	Network devices	Peer to Peer	Data (size)	Application management	IoT	–
Nazmudeen et al. [22]	Network devices	Master-Slave	Data (size)	Data management	PLC	–
Aazam et al. [23]	Network devices	–	Data (size)	Network management	IoT	Data encryption
Cirani et al. [24]	Network devices	–	Context (application)	Network management	IoT	DoS attack
Dsouza et al. [25]	Cloudlets	Peer to Peer	Context (application)	Network management	IoT	Authentication
Cardellini et al. [26]	Cloudlets	Cluster	Context (application)	Application management	IoT	–
Yan et al. [27]	Base stations	Cluster	Context (user)	Application management	RAN	–
Gu et al. [28]	Base stations	Peer to Peer	Cost (deployment, communication)	Resource management	IoT	–
Truong et al. [29]	Base stations	Master-Slave	Context (application)	Network management	Vehicular network	–
Oueis et al. [30]	Base stations	Cluster	Time (deadline)	Latency management	Mobile network	–
Hou et al. [31]	Vehicles	Cluster	Context (user)	Resource management	Vehicular	Privacy
Ye et al. [32]	Vehicles	–	Time (deadline)	Application management	Vehicular	–
Oueis et al. [33]	Base stations	Cluster	Data (flow)	Resource management	Mobile network	–
Faruque et al. [34]	Network devices	Cluster	Energy consumption	Power management	IoT	–

(continued)

Table 1 (continued)

Work	Fog nodes	Nodal collaboration	Provisioning metrics	SLOs	Applicable network	Security concerns
Shi et al. [35]	Network devices	Peer to Peer	Context (application)	Application management	Mobile network	–
Giang et al. [36]	Network devices	Peer to Peer	Data (flow)	Application management	IoT	–
Intharawijitr et al. [37]	Servers	–	Time (communication, computation)	Latency management	Mobile network	–
Peng et al. [38]	Base stations	Peer to Peer	Data (size)	Network management	RAN	–
Hassan et al. [39]	Network devices	Cluster	Cost (execution, communication)	Application management	Mobile network	Privacy
Zhang et al. [40]	Cloudlets	Peer to Peer	Cost (deployment)	Cost management	LRPON	–
Deng et al. [41]	Network devices	Peer to Peer	Data (Size)	Application management	Mobile network	–
Do et al. [42]	Network devices	–	Energy consumption	CO_2 management	CDN	–
Aazam et al. [43]	Servers	–	Context (user)	Resource management	IoT	–
Datta et al. [44]	Network devices	Peer to Peer	Context (user)	Data management	Vehicular network	–
Aazam et al. [45]	Servers	–	Context (user)	Resource management	IoT	–
Gazis et al. [46]	Network devices	–	Context (application)	Resource management	IoT	–

resource management etc. On the other hand, reliability in Fog computing can be discussed in terms of consistency of Fog nods, availability of high performance services, secured interactions, fault tolerance etc. In the existing literature a very narrow discussion towards sustainable and reliable Fog computing has been provided. Further research in this area is highly recommended for the desired performance of Fog computing.

11.3 Interoperable Architecture of Fog Nodes

Generally, Fog nodes are specialised networking components with computational facilities. More precisely, besides performing traditional networking activities like packet forwarding, routing, switching, etc., Fog nodes perform computational tasks. In some scenarios where real time interactions are associated, Fog nodes have to perform more as a computational component rather than a networking component. In other cases, networking capabilities of Fog nodes become prominent over computational capabilities. Therefore, an interoperable architecture of Fog nodes that can be self customized according to the requirements is very necessary. In existing literature although many unique Fog nodes architecture have been proposed, the real interoperable architecture of Fog nodes are still required to be investigated.

11.4 Distributed Application Deployment

Fog nodes are distributed across the edge and not all of them are highly resource occupied. In this case, large scale application deployment on single Fog node is not often feasible. Modular development of large scale applications and their distributed deployment over resource constrained Fog nodes can be an effective solution. In existing literature of Fog computing several programming platforms for distributed application development and deployment have been proposed. However, the issues regarding distribute application deployment such as latency management, dataflow management, QoS assurance, edge-centric affinity of real-time applications etc. have not been properly addressed.

11.5 Power Management Within Fog

Fog nodes have to deal with huge number of service requests coming from end devices/sensors simultaneously. One of the trivial solutions is to deploy Fog nodes in the environment according to the demand. However, this approach will increase the number of computationally active Fog nodes to a great extent, that eventually affects total power consumption of the system. Therefore, while responding large number service requests, proper power management within Fog network is very necessary. However in existing literature, Fog computing have been considered for minimizing power consumption in Cloud datacentres. Optimization of energy usage within the Fog network are yet to be investigated. Moreover, in order to manage power in Fog environment, consolidation of Fog nodes by migrating tasks from one node to another node can be effective in some scenarios. Investigation towards the solutions of optimal task migration can also be a potential field of Fog based research.

11.6 Multi-tenant Support in Fog Resources

Available resources of Fog nodes can be virtualized and allocated to multiple users. In the existing literature, multi-tenant support in Fog resources and scheduling the computation tasks according to their QoS requirements have not been investigated in detail. Future researches can be conducted targeting this limitation of existing literature.

11.7 Pricing, Billing in Fog Computing

Fog computing can provide utility services like Cloud computing. In Cloud computing typically users are charged according to the horizontal scale of usage. Unlike Cloud computing, in Fog vertical arrangement of resources contributes to the expenses of both users and providers to a great extent. Therefore, the pricing and billing policies in Fog generally differ significantly from the Cloud oriented policies. Besides, due to lack of proper pricing and billing policies of Fog based services, most often users face difficulty in identifying suitable providers for conducting SLA. In such circumstance, a proper pricing and billing policy of Fog based services will surely be considered as potential contribution in field of Fog computing.

11.8 Tools for Fog Simulation

Real-world testbed for evaluating performance of Fog based policies is often very expensive to develop and not scalable in many cases. Therefore, for preliminary evaluation of proposed Fog computing environments many researchers look for efficient toolkit for Fog simulation. However, till now very less number of Fog simulator are available (e.g. *iFogSim* [47]). Development of new efficient simulator for Fog computing can be taken into account as future research.

11.9 Programming Languages and Standards for Fog

Basically Fog computing has been designed for extending Cloud based services such as IaaS, PaaS, SaaS, etc. to the proximity of IoT devices/sensors. As the structure of Fog differs from Cloud, modification or improvement of existing standards and associate programming languages to enable Cloud-based services in Fog are highly required. Moreover, for seamless and flexible management of large number connections in Fog, development of efficient networking protocols and user interfaces are also necessary.

12 Summary and Conclusions

In this chapter, we surveyed recent developments in Fog computing. Challenges in Fog computing is discussed here in terms of structural, service and security related issues. Based on the identified key challenges and properties, a taxonomy of Fog computing has also been presented. Our taxonomy classifies and analyses the existing works based on their approaches towards addressing the challenges. Moreover, based on the analysis, we proposed some promising research directions that can be pursued in the future.

References

1. Dastjerdi, A., H. Gupta, R. Calheiros, S. Ghosh, and R. Buyya. 2016. Chapter 4—fog computing: Principles, architectures, and applications. In *Internet of Things: Principles and Paradigms*, ed. R. Buyya, and A.V. Dastjerdi, 61–75. New York: Morgan Kaufmann.
2. Sarkar, S., and S. Misra. 2016. Theoretical modelling of fog computing: A green computing paradigm to support iot applications. *IET Networks* 5(2): 23–29.
3. Bonomi, F., R. Milito, J. Zhu, and S.Addepalli. 2012. Fog computing and its role in the internet of things. In *Proceedings of the first edition of the MCC workshop on Mobile cloud computing*, ACM, 13–16.
4. Sarkar, S., S. Chatterjee, and S. Misra. 2015. Assessment of the suitability of fog computing in the context of internet of things. *IEEE Transactions on Cloud Computing* PP(99): 1–1.
5. Garcia Lopez, P., A. Montresor, D. Epema, A. Datta, T. Higashino, A. Iamnitchi, M. Barcellos, P. Felber, and E. Riviere. 2015. Edge-centric computing: Vision and challenges. *ACM SIGCOMM Computer Communication Review* 45(5): 37–42.
6. Varghese, B., N. Wang, S. Barbhuiya, P. Kilpatrick, and D.S. Nikolopoulos. 2016. Challenges and opportunities in edge computing. In *Proceedings of the IEEE International Conference on Smart Cloud*, 20–26.
7. Shi, W., J. Cao, Q. Zhang, Y. Li, and L. Xu. 2016. Edge computing: vision and challenges. *IEEE Internet of Things Journal* 3(5): 637–646.
8. Hu, Y.C., M. Patel, D. Sabella, N. Sprecher, and V. Young. 2015. Mobile edge computinga key technology towards 5g. *ETSI White Paper* 11: 1–16.
9. Klas, G.I. 2015. Fog Computing and Mobile Edge Cloud Gain Momentum Open Fog Consortium, ETSI MEC and Cloudlets. http://yucianga.info/?p=938.
10. Cau, E., M. Corici, P. Bellavista, L. Foschini, G. Carella, A. Edmonds, and T.M. Bohnert. 2016. Efficient exploitation of mobile edge computing for virtualized 5g in epc architectures. In *4th IEEE International Conference on Mobile Cloud Computing, Services, and Engineering (MobileCloud)*, (March 2016), 100–109.
11. Ahmed, A., and E. Ahmed. 2016. A survey on mobile edge computing. In *Proceedings of the 10th IEEE International Conference on Intelligent Systems and Control (ISCO 2016)*, Coimbatore, India.
12. Mahmud, M.R., M. Afrin, M.A. Razzaque, M.M. Hassan, A. Alelaiwi, and M. Alrubaian. 2016. Maximizing quality of experience through context-aware mobile application scheduling in cloudlet infrastructure. *Software: Practice and Experience* 46(11): 1525–1545. spe.2392.
13. Sanaei, Z., S. Abolfazli, A. Gani, and R. Buyya. 2014. Heterogeneity in mobile cloud computing: Taxonomy and open challenges. *IEEE Communications Surveys and Tutorials* 16(1): 369–392.
14. Bahl, P., R.Y. Han, L.E. Li, and M. Satyanarayanan. 2012. Advancing the state of mobile cloud computing. In *Proceedings of the third ACM workshop on Mobile cloud computing and services*, ACM, 21–28.

15. Satyanarayanan, M., G. Lewis, E. Morris, S. Simanta, J. Boleng, and K. Ha. 2013. The role of cloudlets in hostile environments. *IEEE Pervasive Computing* 12(4): 40–49.
16. Lee, W., K. Nam, H.G. Roh, S.H. Kim. 2016. A gateway based fog computing architecture for wireless sensors and actuator networks. In *18th International Conference on Advanced Communication Technology (ICACT)*, IEEE, 210–213.
17. Aazam, M., and E.N. Huh. 2015. Fog computing micro datacenter based dynamic resource estimation and pricing model for iot. *In IEEE 29th International Conference on Advanced Information Networking and Applications*. (March 2015), 687–694.
18. Jalali, F., K. Hinton, R. Ayre, T. Alpcan, and R.S. Tucker. 2016. Fog computing may help to save energy in cloud computing. *IEEE Journal on Selected Areas in Communications* 34(5): 1728–1739.
19. Zhu, J., D.S. Chan, M.S. Prabhu, P. Natarajan, H. Hu, F. Bonomi. 2013. Improving web sites performance using edge servers in fog computing architecture. In *Service Oriented System Engineering (SOSE), 2013 IEEE 7th International Symposium on*, (March 2013), 320–323.
20. Zeng, D., L. Gu, S. Guo, Z. Cheng, and S. Yu. 2016. Joint optimization of task scheduling and image placement in fog computing supported software-defined embedded system. *IEEE Transactions on Computers* PP(99): 1–1.
21. Hong, K., D. Lillethun, U. Ramachandran, B. Ottenwälder, and B. Koldehofe. 2013. Mobile fog: A programming model for large-scale applications on the internet of things. In *Proceedings of the second ACM SIGCOMM workshop on Mobile cloud computing*, ACM, 15–20.
22. Nazmudeen, M.S.H., A.T. Wan, and S.M. Buhari. 2016. Improved throughput for power line communication (plc) for smart meters using fog computing based data aggregation approach. In *IEEE International Smart Cities Conference (ISC2)*, (Sept 2016), 1–4.
23. Aazam, M., and E.N. Huh. 2014. Fog computing and smart gateway based communication for cloud of things. In *Future Internet of Things and Cloud (FiCloud), International Conference on IEEE (2014)*, 464–470.
24. Cirani, S., G. Ferrari, N. Iotti, and M. Picone. 2015. The iot hub: A fog node for seamless management of heterogeneous connected smart objects. In *12th Annual IEEE International Conference on Sensing, Communication, and Networking-Workshops (SECON Workshops), IEEE (2015)*, 1–6.
25. Dsouza, C., G.J. Ahn, and M. Taguinod.2014. Policy-driven security management for fog computing: Preliminary framework and a case study. In: *IEEE 15th International Conference on Information Reuse and Integration (IRI), (Aug 2014)*, 16–23.
26. Cardellini, V., V. Grassi, F.L. Presti, and M. Nardelli. 2015. On qos-aware scheduling of data stream applications over fog computing infrastructures. In *IEEE Symposium on Computers and Communication (ISCC)*, (July 2015), 271–276.
27. Yan, S., M. Peng, and W. Wang. 2016. User access mode selection in fog computing based radio access networks. In *IEEE International Conference on Communications (ICC)*,(May 2016), 1–6.
28. Gu, L., D. Zeng, S. Guo, A. Barnawi, and Y. Xiang. 2015. Cost-efficient resource management in fog computing supported medical cps. *IEEE Transactions on Emerging Topics in Computing* PP(99): 1–1.
29. Truong, N.B., G.M. Lee, and Y. Ghamri-Doudane. 2015. Software defined networking-based vehicular adhoc network with fog computing. In *IFIP/IEEE International Symposium on Integrated Network Management (IM)*,(May 2015), 1202–1207.
30. Oueis, J., E.C. Strinati, S. Sardellitti, and S.Barbarossa. 2015. Small cell clustering for efficient distributed fog computing: A multi-user case. In *Vehicular Technology Conference (VTC Fall)*, IEEE 82nd. (Sept 2015), 1–5.
31. Hou, X., Y. Li, M. Chen, D. Wu, D. Jin, and S. Chen. 2016. Vehicular fog computing: A viewpoint of vehicles as the infrastructures. *IEEE Transactions on Vehicular Technology* 65(6): 3860–3873.
32. Ye, D., M. Wu, S. Tang, and R. Yu. 2016. Scalable fog computing with service offloading in bus networks. In *IEEE 3rd international Conference on Cyber Security and Cloud Computing (CSCloud)*, (June 2016), 247–251.

33. Oueis, J., E.C. Strinati, and S. Barbarossa. 2015. The fog balancing: Load distribution for small cell cloud computing. In *IEEE 81st Vehicular Technology Conference (VTC spring)*, (May 2015), 1–6.
34. Faruque, M.A.A., and K. Vatanparvar. 2016. Energy management-as-a-service over fog computing platform. *IEEE Internet of Things Journal* 3(2): 161–169.
35. Shi, H., N. Chen, and R. Deters. 2015. Combining mobile and fog computing: Using coap to link mobile device clouds with fog computing. In *IEEE International Conference on Data Science and Data Intensive Systems*, (Dec 2015), 564–571.
36. Giang, N.K., M. Blackstock, R. Lea, and V.C.M.Leung. 2015. Developing iot applications in the fog: A distributed dataflow approach. In *5th International Conference on the Internet of Things (IOT)*, (Oct 2015), 155–162.
37. Intharawijitr, K., K. Iida, and H. Koga. 2016. Analysis of fog model considering computing and communication latency in 5g cellular networks. In *IEEE International Conference on Pervasive Computing and Communication Workshops (PerCom Workshops)*, (March 2016), 1–4.
38. Peng, M., S. Yan, K. Zhang, and C. Wang. 2016. Fog-computing-based radio access networks: Issues and challenges. *IEEE Network* 30(4): 46–53.
39. Hassan, M.A., M. Xiao, Q. Wei, and S.Chen. 2015. Help your mobile applications with fog computing. In *12th Annual IEEE International Conference on Sensing, Communication, and Networking - Workshops (SECON Workshops)*, (June 2015), 1–6.
40. Zhang, W., B. Lin, Q. Yin, and T. Zhao. 2016. Infrastructure deployment and optimization of fog network based on microdc and lrpon integration. *Peer-to-Peer Networking and Applications* 1–13.
41. Deng, R., R. Lu, C. Lai, T.H. Luan, and H. Liang. 2016. Optimal workload allocation in fog-cloud computing towards balanced delay and power consumption. *IEEE Internet of Things Journal* PP(99): 1–1.
42. Do, C.T., N.H. Tran, C. Pham, M.G.R. Alam, J.H. Son, and C.S. Hong. 2015. A proximal algorithm for joint resource allocation and minimizing carbon footprint in geo-distributed fog computing. In *International Conference on Information Networking (ICOIN)*, (Jan 2015), 324–329.
43. Aazam, M., M. St-Hilaire, C.H. Lung, and I. Lambadaris. 2016. Pre-fog: Iot trace based probabilistic resource estimation at fog. In *13th IEEE Annual Consumer Communications Networking Conference (CCNC)*, (Jan 2016), 12–17.
44. Datta, S.K., C. Bonnet, and J. Haerri. 2015. Fog computing architecture to enable consumer centric internet of things services. In *International Symposium on Consumer Electronics (ISCE)*, (June 2015), 1–2.
45. Aazam, M., M. St-Hilaire, C.H. Lung, and I. Lambadaris. 2016. Mefore: Qoe based resource estimation at fog to enhance qos in iot. In *23rd International Conference on Telecommunications (ICT)*, (May 2016), 1–5.
46. Gazis, V., A. Leonardi, K. Mathioudakis, K. Sasloglou, P. Kikiras, and R. Sudhaakar. 2015. Components of fog computing in an industrial internet of things context. In *12th Annual IEEE International Conference on Sensing, Communication, and Networking - Workshops (SECON Workshops)*, (June 2015) 1–6.
47. Gupta, H., A.V. Dastjerdi, S.K. Ghosh, and R. Buyya. 2016. ifogsim: A toolkit for modeling and simulation of resource management techniques in internet of things, edge and fog computing environments. arXiv preprint arXiv:1606.02007.

Challenges and Opportunities in Designing Smart Spaces

Yuvraj Sahni, Jiannong Cao and Jiaxing Shen

Abstract In the past decade, research in Internet of Things and related technologies such as Ubiquitous Computing has fueled the development of Smart Spaces. Smart space does not just mean interconnection of different devices in our surroundings but an environment where the devices respond to human behavior and needs. To achieve this vision, services that are based on user's intents and their high-level goals should be provided. However, existing works mostly focus on providing context-awareness based services. In the past, smart space developers focused on providing technology-centric solutions but this approach failed to achieve wider market adoption of products as users either did not want the solutions at first place or they just could not understand how it worked. Therefore, researchers and smart space developers have now shifted towards the user-centric approach for developing smart spaces. It is non-trivial to develop user-centric smart spaces as developers have to consider factors such as user requirements, behavior etc. apart from usual technical challenges. In this work, we take a comprehensive look at the challenges in developing user-centric smart spaces for two different smart space scenarios: Smart Home and Smart Shopping. We give four user-centric criteria to compare these two smart spaces. At the end, we also provide some future research directions for developing Smart Spaces.

Keywords Smart Spaces · User-centric · Smart home · Smart shopping · Internet of Things

Y. Sahni · J. Cao (✉) · J. Shen ·
Department of Computing, The Hong Kong Polytechnic University,
Hung Hom, Hong Kong
e-mail: csjcao@comp.polyu.edu.hk

Y. Sahni
e-mail: csysahni@comp.polyu.edu.hk

J. Shen
e-mail: csjshen@comp.polyu.edu.hk

B. Di Martino et al. (eds.), *Internet of Everything*, Internet of Things,
https://doi.org/10.1007/978-981-10-5861-5_6

1 Introduction

Internet of Things has become more than a marketing buzzword now. Cisco has predicted that global market of Internet of Things will be 14.4 trillion dollars by 2022. Internet of Things envisions a future where all the objects around us will be connected to each other. This vision is shared by many other interrelated research paradigms such as Ubiquitous Computing, Pervasive Computing, Cyber-Physical Systems, wireless sensor networks etc. The objective of all these research areas is to make our lives more comfortable by using devices with communication and computation capability that are connected to each other to sense our surroundings.

However, these areas do not focus much on the emotional and social side of connectivity. This means that the solutions provided by these technologies just strive for providing automation rather than also helping in connecting people with each other. Due to the proliferation of numerous tech gadgets such as smartphones, laptops, smart watches etc. we are beginning to lose touch with our natural surroundings and even alienating us from other people. Therefore, we need technologies that will allow people to be emotionally attached to their surroundings and help in developing the social connection with other people. This implies that we not only need to connect objects in our surroundings with each other but also people with other people and people with other objects. Internet of Everything is based on the same objective of extending networked connection of objects to include people, process, and data. Cisco defines Internet of Everything as intelligent connection of people, process, data, and things that creates new capabilities, richer experiences and unprecedented economic opportunity for business, individuals, and countries [5, 11]. Figure 1 shows the interconnection of people, data, processes and things in Internet of Everything.

Instead of just focusing on technological aspects of an application, researchers are now trying to use knowledge from multiple disciplines such as sociology, psychology, philosophy, architecture etc. to design an application. Researchers want to use

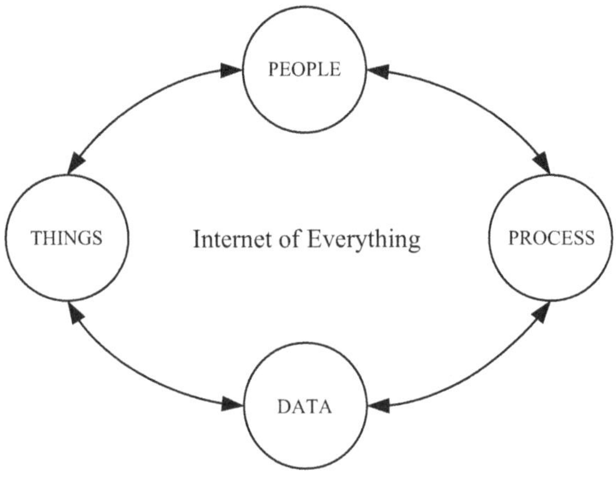

Fig. 1 Interconnection of people, process, data, and things

the knowledge of human emotions, social connections, and interaction between surrounding devices and humans with each other to provide improved services to users. Smart Spaces is one such application which tries to make our surroundings smarter by utilizing the knowledge from multiple disciplines. Many IoT applications such as Smart Home, Smart Building, Smart HealthCare, Smart Parking, Smart Retail etc. can be classified as a type of Smart Space. All of these applications are somehow interconnected as there is sharing of data between each other. No matter what the approach is for designing each application, final objective of each application is to improve user's life by providing better services. Although the specific details might be different but the challenges such as interoperability, scalability, security, privacy, etc. are also common for every application. Since these applications are so closely related, it makes sense to understand them together.

In this paper, we give an overview of Smart Spaces in general and then study in detail about two important applications i.e. Smart Home and Smart Shopping. We look at drawbacks in current solutions and classify the reasons why these applications are not being widely accepted by users. According to our analysis, we found that if smart space developers want to have wider market adoption of their technologies then they should shift their focus from technology-centric view to user-centric. Smart space developers should not compromise on some essential features such as low cost, high security, reliability, flexibility and robustness, and easy manageability to enable wider market adoption. In coming future, all the smart spaces will be combined with each other so, it is important to understand the difference various smart spaces in order to combine them. Therefore, we have also given four user-centric criterias (type of stakeholders, number of users, dynamicity of smart space, and user's requirement) to compare smart home and smart shopping application. After analyzing the challenges and drawbacks in smart spaces, we also provide some future research directions for developing smart spaces.

The rest of the paper is as follows. Section 2 gives a generic overview of Smart spaces. Sections 3 and 4 discuss in detail about Smart Home and Smart Shopping application respectively. Section 5 discusses the difference between Smart Home and Smart Shopping. Finally, in Sect. 6 some research directions for developing smart spaces are provided.

2 Overview of Smart Spaces

Smart Space is any surrounding environment that adapts itself to human behavior and needs by utilizing the data obtained from the interaction between objects and humans. The "objects" here refer to all the devices that are present in our surrounding which may include wearables, smartphones, laptops, or any other device capable of sensing and/or actuation. The objects and users within a smart space can be either stationary or mobile. By using the data from various social networks and other devices in the surroundings, we can analyze and obtain the contextual information and data related to user behavior and requirements. Once we know the user requirements we can use it to provide personalized services and make the lives of users more comfortable.

The development of smart spaces requires knowledge from multiple disciplines such as computer science, psychology, sociology, architecture etc. We need to collect data from sensors and other sources, analyze this data to find some useful features related to human behavior, exchange this data with heterogeneous devices and then configure the devices and systems accordingly. Interactive user interfaces are also one of the most important components to be included in smart space as they make it easier to manage the smart spaces. User-friendly interfaces are required to display the result obtained from different sources of data and enable the interaction with different devices and systems. These interfaces also open new opportunities for exchanging data among users and enable better collaboration among individuals.

Technical challenges such as interoperability, resource discovery, scalability, big data analytics, openness, robustness, security, and privacy are common for every smart space scenario [48]. Interoperability is a major research challenge that needs to be resolved to allow interaction between devices or users located within and across different smart spaces. European Research Cluster on the Internet of Things (IERC) defines four types of interoperability i.e. technical, syntactical, semantic, and organizational interoperability [51]. Technical interoperability is related to hardware/software components and communication protocols that enable machine to machine communication. Syntactical and semantic are related to format, syntax, and meaning of data. Organizational interoperability is about overall ability to communicate and exchange data between two different organizations. Smart spaces need to support the capability to add new devices, users to the existing system and also allow different smart spaces to exchange data with each other. Smart Space is a dynamic environment that consists of a large number of devices and users interacting with each other. Some of the scenarios that need to be handled while managing a smart space are:

- Addition or removal of devices: Since all the devices interact with each other to provide a comfortable environment, addition or removal of a device will at least require informing the other devices about the change in the configuration of network. Addition or removal of devices will lead to changes in the connectivity and coverage of the network. There is a possibility that addition of new device may make an old device redundant or outdated so the old device would have to be removed. On the other hand, if any functionality was being commonly handled by the removed device and another device, then the other device will have to change its configuration accordingly.
- Changing the configuration of a device: A device configuration could be changed with time. This change could be either with hardware or software. This change might make some devices incompatible for data exchange which will hamper the functionality of the whole system. Therefore, changes in one device will reflect in all the network and other devices will have to configure themselves accordingly.
- Reconfiguring the Smart Space according to the user: Nowadays, the services being provided are usually personalized. Each user has different preferences and

therefore the user has to modify the settings of devices according to his/her requirements. This problem can be resolved if the smart space can recognize the user and remember the users' settings. So the next time if the same user enters the smart space, device settings are changed automatically [10].

- Handling multiple users' requirement simultaneously: In the previous point, we made the assumption that there is only user present in the smart space. But usually, within a home building or office, there are multiple individuals that are present at any single time. Since each user might have different preferences, it is very difficult to adapt the smart space such that it is suitable for every user. This is an ongoing research challenge to resolve the conflict arising due to multiple users' requirement [39].

Although the technical issues are important in developing smart spaces but if the researchers want their technological solutions to be widely used by everyone they need to change their approach. Therefore, in recent years, researchers have changed their approach from technology-centric to user-centric. Researchers are focusing more on the requirements of users rather than just thinking about the new technological solutions they can provide. Previous method of just pushing technology into the market did not work so well as users either did not want the solutions at first place or they just could not understand how it worked. We have outlined some of the non-technical issues below that need to be taken into consideration while developing smart spaces.

1. User Profile: It is important to understand whether the smart space is intended to be used by a specific set of users or the solutions provided are applicable for everyone [3]. For e.g.: Ambient assisted living is a smart space application that is usually designed for elderly people and it has to be different from smart space that is designed especially for young kids. This example illustrates the difference in age but in fact, users could be different in terms of habits, social needs, physical and mental health etc.
2. User's Knowledge about Smart Space: Usually an average user has very little understanding of what is smart space, what are the functions of different devices, and how to configure those devices according to their requirements. In [38, 55], the experience of users operating smart devices in a natural home environment has been studied and it was observed that users cannot fully understand the system behavior so they have to try some hacks to configure the system settings. This kind of situation leads to user frustration.
3. User-Device Interaction: User interfaces for devices within smart spaces must be interactive, simple to use, require low effort for understanding, and most importantly usable by all kinds of users [3]. Yang and Newman [55] analyzed the use of Nest thermostat in natural home settings, it was revealed that good interface design leads to better engagement. Researchers have tried various types of interfaces such as gestures, audio-visual, brain-computer interface. Nowadays, researchers are trying to create interfaces that enable people to interact with their natural surroundings. For e.g. in [25] an interactive interface called "time home pub" has been designed that uses table, whiskey glass, MP3 player as main components for interacting with surroundings.

4. Balance between User-device control: It is important to decide how much control should be given in the hands of users. We could either have a case where users directly control the space around them or another scenario where devices passively monitor the users' behavior and needs and then configure the space accordingly. It has been found that if users feel out of control or do not understand the working of devices while using autonomous technologies then they impose limitations on the level of automation [39]. This means they might set the settings of a device manually rather than depending on it. Mennicken et al. [39] suggests that it is better to consider in terms of collaboration between users and devices rather than control. In this case, both user and devices exchange useful information with each other in order to make any decision.

3 Smart Home

Smart home is a residential area that automatically adapts itself according to resident's requirements and allows them to access and control their surroundings that are being monitored using various sensors and other devices. Various kinds of sensors embedded in wearables, smartphones, and surrounding devices collect data related to physical environment, human behavior, and human activities. This data is then analyzed to automatically adapt the physical environment and provide a range of personalized services to humans that help in improving their living experience [13]. Different individuals use smart home services for various objectives but we can classify them into four main types as shown in Fig. 2.

According to a study done in [35], average US citizen spends 15.6 hours inside a home. Since this is almost 2/3rd of our daily time, it becomes essential to pro-

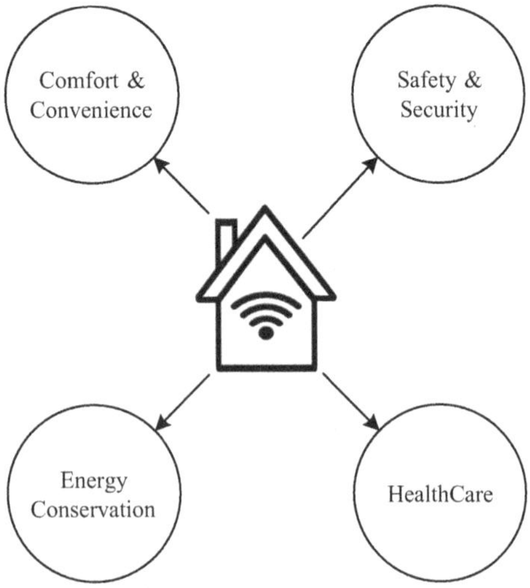

Fig. 2 Classification of smart home services

vide functionalities that can enhance our comfort while staying inside a home. These functionalities can include providing remote access and control of various appliances within a home, automatically adapting HVAC systems according to physical environment and other contextual information, providing improved security by allowing access to authorized individuals, monitoring the health conditions of inhabitants and sending an alert in case of abnormal situation (fall detection, heart attack etc.), or setting the entertainment systems according to your emotions [4, 54]. In reference [42], authors provide a list of twenty-two services such as smart memories, smart bed, smart table, smart bathroom, smart wardrobe etc. that can be included in a smart home. Authors in [50] use computing technologies to transform normal surfaces inside a home such as a fridge door, kitchen walls, notice boards into smart surfaces that can help us in efficiently organizing our home life. IEEE has created a virtual home, IoT Home of the future, that shows the technologies and functionalities that can be included in a smart home in coming future [26]. Researchers have also created real smart homes such as Mavhome [15], Georgia tech aware home [29], House_n [49] to demonstrate the possible functionalities that could be included in future smart homes.

Most of the services developed for smart home try to enhance our comfort level. Even though comfort and convenience are a priority while developing smart homes, we cannot ignore the damage that could be done to our natural surroundings by over-utilizing the resources like energy. Therefore, there is always a debate between comfort vs energy i.e. whether we should prefer energy-conserving environment or use functionalities that maximize our comfort [39]. This leads to another point of view for smart homes where the focus is on saving energy and money by utilizing energy management systems that also help in reducing the carbon footprint [54]. The basic idea is to use smart meters and other interfaces that inform the user about the total energy being consumed and provide possible solutions that will help in saving the power and money for inhabitants. Energy management systems can be used to program (either automatically or manually) the appliances inside the home such that they are not used at the time of peak electricity price, and they get switched off when not in use or when total power consumption exceeds a threshold. These settings are dependent on the kind of household and their energy demands.

Out of all the smart home applications, ambient assisted living (AAL) has received the most attention by researchers working in this area. AAL aims to make the lives of people with special demands such as elderly, handicapped etc. more comfortable by enabling them to live independently at home [30]. Factors such as increasing aging population, high cost of professional health care personnel, increasing burden on professional health care personnel, and increasing demand of people to continue living independently at their current place of residence has prompted researchers to put more emphasis on this application [46]. It is very challenging to provide a comfortable life for elderly as they generally face issues like the decline in physical activity, vision, hearing, cognitive functionality, and even many age-related diseases such as Alzheimer, Parkinson, Arthritis etc. [46]. Some of the important techniques required for helping the elderly and other such individuals are human activity recognition (to detect daily life patterns) [14], planning (to help plan activities especially for patients suffering from dementia), anomaly detection (to detect wandering patterns or hazardous behavior) [12, 17], identity detection [21] and indoor localization (to track and provide location based services), context modeling (to provide context based services) etc. [46].

While designing solutions for AAL, researchers should take into account the special requirements of the specific individual and continuously monitor whether their current situation or illness affects their capability to use provided technology [23]. According to a study done in [23], it is seen that these individuals, especially elderly, care about connecting and communicating with their peers and other family members. Other important finding from studies done in [4, 23, 30, 46] is that elderly people do not accept modern IT technologies easily. There is also a social stigma attached to using these solutions that it makes them look dependent and in need of professional health care [23]. So they often try to hide the wearables or other sensory devices in their surroundings. Elderly people need technologies that are unobtrusive and adaptable according to specific individuals and context [30].

In recent years, researchers have come up with many innovative solutions that help in solving issues related to AAL. In [33], authors propose some guidelines in adapting the prompting strategies (auditory, pictorial, video or light) according to the cognitive profile of the patients suffering from Alzheimer's Disease. Since privacy and unobtrusiveness is an important concern for individuals [13], authors in [1] implement a device called vital radio that uses reflection of low power wireless signals off human body to track breathing without violating privacy or using any contact with human body. The technology has reached a point where we can even help in saving a life. Authors in [6] show a case study where it is revealed that life of a patient could have been saved from heart attack by analyzing real-time data from combination of multiple sources such as changes in activities, data from body worn and surrounding sensors, data from medical devices etc.

Apart from AAL application, we have plethora of smart home devices emerging in the market. Every major company including Google, Microsoft, Samsung, Apple, Amazon etc. are introducing devices that promise to automatically adapt our surroundings and make our homes smarter. According to report by IControl Networks that surveyed 1600 consumers [41], 90% of consumers purchase smart home products for increased personal and home security, 70% for saving energy and money, and entertainment being the new emerging factor for buying smart home products. Another interesting trend observed is that 60% people prefer devices that can adapt themselves automatically. It shows that people are ready for smart homes, however, the adoption of the smart home devices is still very low. In [55], study was done to determine problems faced by residents using intelligent systems like NEST thermostat. It was revealed in [55] that users face problem understanding the learning behavior of NEST and in some cases users were even annoyed by the adaptive changes done by Nest. This issue leads to users taking over the control of devices instead of relying on automation done by devices. We identified four major reasons behind low acceptance of smart home products by users which are lack of consideration of user profile, high cost, high complexity, and lack of trust. Each of these issues has been explained in detail in the following subsections.

3.1 Lack of Consideration of User Profile

As mentioned before, most of the research in smart home has been focused towards health related users and even then it is an ongoing research challenge to determine the user attributes for designing home health care technologies [9]. As for other types of users, a lot of research is required to obtain specific and differentiating characteristics [54]. Users differ in terms of age, gender, profession, socio-cultural beliefs, acceptance of technology, physical and mental health, social needs, daily routine, social relationships etc. An individual also changes with time, so a smart home system that works now may not work in near future due to change in user with time [39]. Looking at these differences, it is apparent that designing a smart home even for a single person is very challenging as it needs to be very flexible and meet such varied demands. Usually, smart home consist of multiple individuals that share the space and devices with each other so the chances of conflict are much higher as each individual has its own preference. We have described four criteria below that will help in determining the type of users and the solutions they prefer.

Diversity of users based on age
Most of the smart home services are designed for people who have been staying in their homes for long time [4]. Even though young people have more acceptance towards technology, they cannot take full advantage of these services because most young people prefer to live in rented homes due to affordability factor and their choice of living. According to PwC, 60% of population will live in rented homes in London [43] therefore, the smart home services need to be made more flexible and cheaper. Young people need smart home services that are modular and independent so that they can use these services even in their new homes without worrying about integration issue. Next group of users belongs to the category of families having children. Apart from affordability and flexibility, this group of users is also concerned with energy savings, and security of their home and people inside it. They are interested in services that can help them in monitoring the activities of their children or to get the energy and cost information. The third category of users is older age people who usually live alone in their homes. One important challenge regarding elderly people is that they do not easily accept new technology. So technological solutions that use smartphones or new gadgets might not be the best choice for them as they may not know how to operate that and are not very eager to learn new technologies [4].

Physical and Mental Health
A smart home solution that is suitable for an average individual will definitely not work for someone who is suffering from an illness or physical disability. Users with special needs have different types and stages of illness so they need solutions that are suited according to their individual context [9]. Authors in [33] show how different patients suffering from Alzheimer's Disease need different prompting strategies according to their cognitive profile. So, even though two individuals may suffer from the same disease, their stage and experience will determine what kind of solution is best suited for them.

Attitude towards smart home automation
Most of the users believe that automating the functionalities in the house will lead to peace of mind and convenience for them [8]. However, everyone does not share the same view as there are some group of users who think that automating functionalities inside the house will make them lazy or they will lose control of their own house [4]. Different users have different philosophical beliefs and cultural differences which makes it difficult to provide a solution that can work for everyone. For example, affluent people who can afford the smart home solutions usually prefer comfort while middle and lower class families want to save money and energy. Another class of users is technophiles who have positive attitude towards adoption of technologies. In recent years, do-it-yourself (DIY) technologies have emerged that allow users to program the smart home solutions themselves. Such solutions are good for technophiles but average user will not adopt them easily as they have very minimal understanding of smart home technologies.

3.2 *High Cost*

Even if a smart home solution meets the demand of an individual, it never comes at a low cost. Cost here is associated with both time and money. Current smart home solutions are expensive which is the major reason behind limited market adoption. Most smart home systems are outsourced and they are not affordable for average households. Users can have cheaper systems by utilizing do-it-yourself (DIY) technologies that also offer more flexibility but user needs to have sufficient technical knowledge to use them and they have to devote lot of time [52]. Another issue with current smart home systems is that they require some structural changes in the house which again costs money and time [8]. People who stay at rented houses cannot afford to make these structural changes so they usually do not adopt them. In coming future, more people will live in rented houses so these issues need to be resolved to allow more adoption of smart home solutions [43].

3.3 *High Complexity*

Users want to adopt smart home solutions to make their life more comfortable and convenient, however, if the solutions are complex for them to understand then they will be more annoyed than comfortable [8]. Users want solutions that can be easily managed and controlled. Interactive interface plays a major role in allowing users to achieve this objective. The interface should be simple enough to be understandable by any user irrespective of age or technical background. A study of experiences of users using home automation technologies was done in [8] and it revealed that users did not like that they had to explain the working of smart devices to anyone new to the home. Authors in [31] design context based notification system that is efficient and less disruptive than traditional notifications by smartphone. Such systems make it easier to view and control the devices. It is often observed that smart home devices are usually managed by just one person in the house who is most likely a technophile or one of the elder member. One of the main objectives of smart home technologies is to improve social connection and emotionally connect users to their surroundings and this is definitely not achieved in the current scenario. In [20], authors propose a game-based collaborative system that uses gamification mechanisms such as points, levels etc. to engage all members in a house to collaboratively manage the devices [20]. Another complaint that is received by smart home users is that they cannot customize their systems and thus they have no control over their own houses. Although DIY technologies do help in customizing the houses but they cannot be used by everyone [52]. Smart home users cannot understand the learning process of devices which is frustrating for them as they think they are not in control [55]. This situation is made worse by the fact that sometimes smart home devices do not respond or function in an undesired manner. They always need the help of outsider or someone with technical knowledge in the house to control these devices [8]. Repair is another issue that creates a problem for smart home users. The systems are so complex for them that they require the help of consultants to do even minor repair or changes in configuration [8].

3.4 Lack of Trust

If the users do not trust the smart home solutions then no matter how smart the solutions are, they will not be adopted. Data collected by sensors in the smart home contains a lot of personal information such as location, behavioral data, daily routines etc. which should be kept private and secure. Smart homes are designed to provide remote access and control to individuals which is appealing to users but if the system is not secure then people with evil motives can use it to their advantage. Hackers can remotely use the system to manipulate our physical environment. Therefore, it is important that devices in the smart home can only be used by authorized individuals [13]. Another important point to consider is to keep the data confidential so that privacy of users is maintained. The third factor that leads to lack of trust among smart home users is unreliability of devices. Smart home users often face situations where the devices start adapting in an undesired manner or they become unresponsive [8, 55]. In future smart homes, devices will make autonomic decisions based on learning the human behavior and sometimes this might lead to undesired behavior. Authors in [18] use the concept of autonomic computing to resolve misunderstanding situations that may arise in futuristic home scenarios.

4 Smart Shopping

Over the last decades, the advances of pervasive computing and data analytics are increasingly transforming regular shopping malls into another smart space, where customers' shopping behaviors can be captured and analyzed, and thus lead to a more user-friendly shopping environment. According to the research results in [47], smart shopping is to minimize the expenditure of time, money, or energy to gain hedonic or utilitarian value from the shopping experience.

There are two aspects, user-oriented and shop-oriented, in smart shopping. Most of current works focus on users' aspect, which can also be classified into two categories. The first category is to understand customers' shopping behaviors; the other category is to enhance customers' shopping experience. Detailed classification is illustrated in Fig. 3.

4.1 User-Oriented Smart Shopping

4.1.1 Enhance Shopping Experience

Brick and Mortar stores have been facing unrelenting competition from online retailers. An enhanced shopping experience is often perceived as a decisive factor in regaining market share. A lot of research efforts have been put into this perspective.

Wang et al. in [53] modeled retail transaction data for personalized shopping recommendation. While an integrated approach for cost-effective development of innovative in-shop-experience applications leveraging the Internet of Things, HTML5 and Pervasive Display Networks is proposed in [37]. Mahashweta et al. [16] proposed a novel recommender system that helps users in shopping for technical products. The

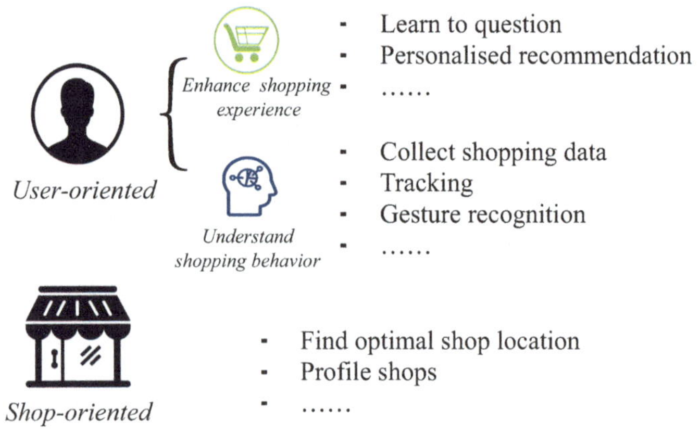

Fig. 3 Classification of smart shopping

suggestions are generated by leveraging both user preferences and technical product attributes. WeShop [32] is a mobile application which uses social data to help customers navigate the decision process in the store. The authors found that uncertainty about a product can act as a barrier to purchase for a customer. The more confident a customer is about a product, the more likely he or she is to purchase it. At the core of the experience is the use of social profile data as a form of context to provide a tailored experience aimed at reducing customer uncertainty.

4.1.2 Understand Shopping Behavior

Retailers are dying to know more about their customers and have a better understanding of customers' shopping behaviors which is critical for market adoption and product promotion. Existing works mostly focus on how to collect customers' shopping data, tracking, and recognize their gestures.

For data collection, TagBooth [36] is an innovative system to detect commodities motion and further discover customers' behaviors, using COTS RFID devices. The authors exploited the motion of tagged commodities by leveraging physical-layer information, like phase and RSS, and then recognize customers' actions like picking, toggling events. Another work is a real-time data collection system proposed in [56], which is based on the following queries.

- To discover the path of a given length (defined by the number of sectors) shared by the largest portion of buyers.
- To find out the path with as many sectors as possible, subject to a predefined threshold of support.
- To find out sectors where buyers visit frequently but seldom purchase any products in these sectors.

For tracking customers, Harikrishna et al. proposed a video analytics solution for tracking customer locations in retail shopping malls [45]. In the work, they presented a computer vision based system for tracking customer locations by recognizing individual shopping carts inside shopping malls in order to facilitate location based services. Customers' traces offer researcher insights about their behaviors. Toshikazu [28] proposed a concept of KANSEI modeling from the aspects of users needs in information service. The key issue is to computationally describe human information processing process from the following aspects; (1) intuitive perception process, (2) subjective interpretation of their situations, (3) knowledge structure of service domain, (4) feature of behavior pattern, and (5) decision making process. Figure 4 illustrates the schematic model of KANSEI.

SangJeong Lee et al. presented a customer malling behavior modeling framework for an urban shopping mall in [34]. The framework utilizes customers' smartphones to derive a holistic understanding of customer behaviors from physical movement to service semantics and proposed a multi-level structure of customer behavior model as shown in Fig. 5.

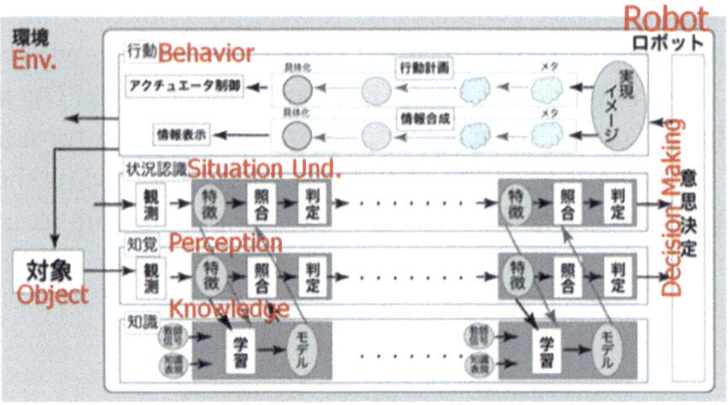

Fig. 4 Schematic model of KANSEI [28]

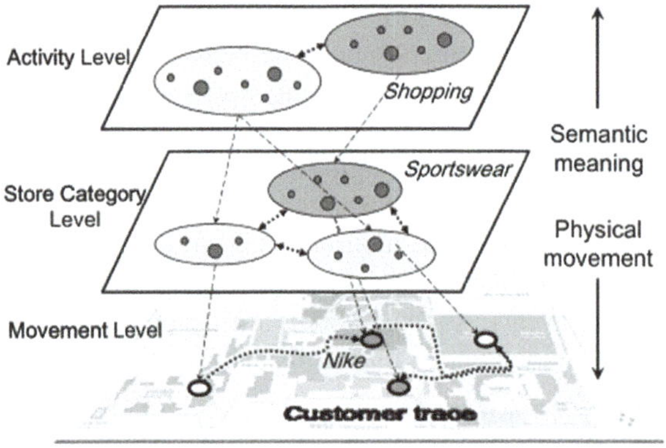

Fig. 5 Multi-level structure of customer behavior model [34]

For recognizing customers' gestures, some researcher used WiFi to sense customers' behaviors in a retail store, since video surveillance can not be used due to high cost and privacy concerns. Zeng et al. [57] showed that various states of a customer such as standing near the entrance to view a promotion or walking quickly to proceed towards the intended item can be accurately classified by profiling Channel State Information (CSI) of WiFi. Also Meera et al. [44] demonstrated that reliably inferring customers' in-store interactions and behaviors by just observing their hand and foot movement inside a store. The hand gestures and locomotive pattern of the customer is identified by appropriately mining the sensor data from shoppers personal smartphone and wearable devices (like smart watches).

4.2 Shop-Oriented Smart Shopping

Numerous research focus on user aspects, only a few of them try to model shops. ShopProfiler [24] is a shop profiling system on crowdsourcing data. First, they extracted movement patterns from customer trajectories. Then localized shops through WiFi heat map. And lastly they categorized shops by designing an SVM classifier in shop space to support multi-label classification and infer brand name from SSID by applying string similarity measurement.

Karamshuk et al. used a data-driven approach to find the optimal location for a new retail store in [27]. They exploited check-in data from Foursquare and mined two features to predict the popularity of retail stores. The two general signals are geographic, where features are formulated according to the types and density of nearby places, and user mobility, which includes transitions between venues or the incoming flow of mobile users from distant areas.

4.3 Immature Techniques

Smart shopping is not that prevalent currently, as some fundamental techniques are immature and cannot be applied to large real scenarios. For example, accurate indoor positioning system require specialized equipments. Cheap as WiFi-based localization systems are, they can only derive coarse-grained location information. Another example is CSI-based gesture recognition. CSI is utilized to recognize customers' gestures, but it does not work when there is a lot of customers, which poses a strong assumption against reality.

5 Discussion

Researchers are trying to make everything in our surroundings smart by introducing a different variety of sensors and devices but currently different smart spaces do not really interact with each other. Our needs and behavior are influenced by every small thing that we interact with in our surrounding. This includes all the devices and people at our home, office or any other place. Therefore, if we want to implement a true "Smart" system, then we need to use data from multiple smart spaces. Different smart spaces not only need to share data but interact with each other. We give an imaginary scenario below where three different smart spaces (Smart Home, Smart Parking, and Smart Shopping) interact with each other. This scenario shows how our life will become more comfortable if multiple smart spaces can share the data and interact with each other. Interaction of different smart spaces will drastically change our way of living.

Let's say there is a scenario where you take your car and go towards Shopping mall to buy some clothes for an upcoming party. Smart Parking application will monitor your trajectory and calculate the time to destination. Based on your previous preference, a parking spot will be reserved for you at the shopping mall and smart parking application will guide you to that particular spot once you reach your destination. At the same time, sensors in your smart home monitor and predict your future requirements. Wearable sensors and sensors on your smartphone analyze your current situation and since you are at a shopping mall, you get a notification that you

Table 1 Difference between smart home and smart shopping application

Difference criteria	Smart home	Smart shopping
Type of stakeholders	1(Household inhabitants)	2 (Shop owner and customer)
Number of users	Less than 10	Greater than 1000 per week
Dynamicity of smart space	Low	High
User's requirement	Personalized surroundings	Personalized recommendation

might need to buy some grocery items as they are almost finished. You select this notification and you get a detailed list of items that need to be bought. Within the shopping mall, smart shopping application will give you personalized recommendations and guide you to make your shopping experience more efficient and enjoyable.

In the coming future, not just these three applications but all the smart spaces that one can imagine such as home, office, hospital, shopping mall, parking lot etc. will interact with each other. There are three main technical challenges that need to be tackled to develop such an integrated system. First one is interoperability to allow sharing of data between heterogeneous systems. Second is scalability so that system is robust enough to add and remove devices/users. Finally, security and privacy cannot be ignored as the interaction of different smart spaces will require access to personal information that should be kept secure.

We analyzed two important smart spaces, Smart Home and Smart Shopping, independently in Sects. 3 and 4 respectively. However, as stated above, we need to think in terms of whole integrated systems rather than individual smart spaces. Even though most of the technical challenges are common for these two smart spaces there are many small differences that should be considered while designing them. We have outlined four main differences (Table 1) below between Smart Home and Smart Shopping application. The four differences given below can also be utilized to differentiate other applications.

1. *Type of Stakeholder*: While developing any technological solution for a smart space, we need to consider who will use the technological solution and what are their requirements. Users who are interested in the smart space solutions are called stakeholders. For any smart home application, we have just one type of stakeholder i.e. household inhabitants. However, these household inhabitants can be further classified into many categories such as children, young people, families, elderly, physically disabled individuals, mentally disabled individuals etc. In Sect. 3, we classified objectives of smart home users into four categories which are comfort and convenience, security, energy conservation, and healthcare. On the other hand, for a smart shopping application, we have two type of stakeholders: i.e. Shop owners and customers. Shop owners are interested in increasing their sales so they want to know different marketing strategies and other useful information that will help them in attracting more customers. While customers want to get the best value for their money and a personalized experience while shopping. Customers are also interested to know the latest update on their favorite products

that are launched into the market. Use of technology can help achieve the objective of both the stakeholders but it is important that these solutions are unobtrusive for customers.

2. *Number of Users*: Scalability is an ongoing research challenge in developing smart spaces. The number of users in a smart home is in the order of tens at maximum while for a smart shopping scenario this number is definitely larger. For super stores like Walmart, this number is around 100,000,000 customers per week [7]. According to Gartner, by the year 2022 number of devices within a single home could be 500 [22]. Currently, we do not have an exact number of devices for smart shopping application but if the number of customers is any indication then the number of devices should at least be in the range of thousands for stores like Walmart. With such huge difference in the number of users and devices for these applications, it is clear that a solution for a smart home cannot be directly applied for smart shopping application.
3. *Dynamicity of Smart Spaces*: Configuration of a smart space can be changed by addition, removal, or change of devices or users in the system. A smart space should be robust enough to recover from any change in its current configuration. Difficulty in developing a smart space directly depends on how dynamic it is. Smart home application is not as dynamic as Smart Shopping. In the case of a smart home, once the systems are configured according to user's requirement they are seldom changed later on. Few changes are done when devices are replaced or new user is added but these changes are minimal. However, for a smart shopping application, there is always a constant change in the number of users. The mobility of users in smart shopping scenario is also higher as compared to smart home scenario. There are higher chances of device damage in smart shopping application as the number and types of users utilizing the devices is higher.
4. *User's Requirement*: Smart Home users want their surroundings to adapt according to their behavior and requirements. For example automatic adaption of lighting or HVAC system within a home. This is called personalized setting of smart home environment. Now if a smart home consists of multiple inhabitants then everyone wants to set the devices according to their own choice which leads to conflict. In case of Smart Shopping scenario, such a conflict does not occur as users are not interested in personalizing the surrounding environment. Customers in smart shopping application are interested in receiving personalized recommendation for shopping. Shop owners collect data related to their customers and use it for personalized marketing of products. In both cases, users want personalized services but the type of service required is entirely different. Smart space developers should consider type of user's requirement while integrating multiple smart spaces.

6 Future Directions for Research in Smart Spaces

Today we have tons of products in the market that are being branded as "Smart" devices. However, when these "Smart" devices are used in a practical environment they do not meet the expectations of users [39]. This is why researchers are now testing their solutions in real situations instead of laboratory settings. In previous sections, we analyzed the drawbacks in Smart Home and Smart Shopping application and even compared these two applications. This section points out some research directions for smart space developers. As it has been mentioned earlier that smart space development requires effort from multiple disciplines so we do not cover all possible research directions. Many issues such as policy-making, legal, ethical, philosophical etc. have not been considered in this section.

1. *Improved Sensing Technology*: Sensing is the fundamental towards development of smart spaces. We use a wide variety of sensors to monitor our physical environments, activities, health signs, and for many other purposes. Authors in [19] classify sensing devices being used in the smart home into three categories i.e. Wearable devices, Direct environment components, and infrastructure mediated system. If we want everything around us to be smarter then we need sensors that have lesser weight, smaller size, and longer battery power and transmission range. Energy harvesting could be a solution to low battery issue but current solutions are not sufficient. Research efforts are required to develop new ways of sensing that are more comfortable and less obtrusive [1]. Issues like absorption of electromagnetic energy by human tissue will be an important concern in coming future as the number of sensing devices around us will be very large [46].
2. *Beyond Human Activity Recognition*: Usually the services provided to users in a smart space are based on the current context and situation. Context and situation awareness is done based on the recognition and prediction of human activities from sensor data [14]. This is not sufficient though because a smart space means the surrounding environment is adapted based on user's behavior and requirements. Therefore, researchers should work towards recognition of high-level goal or intent of users [39]. Research is required to develop new algorithms that can predict human emotions, behavior, comfort and eventually their intent in a naturalistic environment. Another area that needs attention is recognition and prediction of critical events based on collected sensor data [14]. This is important because users are more interested to know about anomalies and critical events rather than regular events [14, 39].
3. *Interactive Interfaces*: Designing interfaces for human-device interaction will continue to be an important issue in coming future. One interesting topic in this research area is to design interfaces for elderly and physical or mentally disabled individuals. Interfaces for these special individuals should be designed differently. One of the major reason for the limited adoption of smart space solutions especially among these individuals is the social stigma attached to using special care facilities [46]. Therefore, they need interfaces that are not only easier to use but also they look more natural and hence are invisible. Interfaces should be designed

such that they can be used by anyone irrespective of their technical background or any other difference. Even though devices are being made to autonomously adapt themselves, humans will still be somehow involved in decision-making process. Future interfaces should be designed not only to allow management of devices but also enable collaboration between devices and humans.

4. *Interoperability*: Use of heterogeneous devices is common for developing a smart space. There are solutions available to handle technical interoperability challenge that occurs due to the difference in communication protocol and standard being used. However, in coming future, we will have multiple smart spaces interacting and sharing data with each other. This means we need interoperability solutions not only to allow transmission of data between completely different systems but also to understand the data being transmitted so that decision-making can be done based on the shared data. Semantic and organizational interoperability will continue to be major challenge at least in coming future [40, 51]. Research efforts are required to develop a standardized architecture for developing smart spaces.
5. *Robustness*: A smart space is a dynamic environment where users come in or go out, and the behavior and requirement of any particular user changes with space and time. Even the devices in a smart space can be added, removed, or changed based on requirement. Both devices and users could be mobile or static at any time. Basically, the condition of both users and devices changes with time. In coming future, the systems will become even more complex so research is required to develop systems that are flexible and robust enough to adapt to such dynamicity. If any system is not robust then it is not reliable for the user to use it. Failure of systems such as fire-alert or other safety system installed in a building could also be life threatening for user [48].
6. *Security and Privacy*: Systems in coming future will support autonomous adaption feature which means they will have data related to user behavior and requirements. Such personal data should not be allowed to fall into the hands of unauthorized entities. Therefore, it is important to address issues such as data authentication, data integrity, data confidentiality etc. In order to protect the privacy of users, researchers have proposed that users should have control over which data is being collected, who is using it and where is it being stored [2]. This solution may not work in coming future as we will have sensors everywhere around us collecting data and since multiple smart spaces will be combined, it will be difficult to have control over who will use it and how. New innovative solutions are required that can address security and privacy issues even for complex and scalable smart spaces that will be developed in coming future.

Acknowledgements The work described in this paper was partially supported by the funding for Project of Strategic Importance provided by The Hong Kong Polytechnic University (Project Code: 1-ZE26), and NSFC project (Project Code: 61332004).

References

1. Adib, F., Z. Kabelac, H. Mao, D. Katabi, and R.C. Miller. 2014. Demo: Real-time breath monitoring using wireless signals. In *Proceedings of the 20th annual international conference on mobile computing and networking*, ACM, 261–262.
2. Atzori, L., A. Iera, and G. Morabito. 2010. The Internet of Things: A survey. *Computer Networks* 54(15): 2787–2805.
3. Balandin, S., and H. Waris. 2009. Key properties in the development of smart spaces. In *International conference on universal access in human-computer interaction*, 3–12, Springer.
4. Balta-Ozkan, N., R. Davidson, M. Bicket, and L. Whitmarsh. 2013. Social barriers to the adoption of smart homes. *Energy Policy* 63: 363–374.
5. Bojanova, I., G. Hurlburt, and J. Voas. 2013. Today, the internet of things. Tomorrow, the Internet of everything. Beyond that, perhaps, the Internet of anything—a radically super-connected ecosystem where questions about security, trust, and control assume entirely new dimensions. *Information-Development*: 04.
6. Bradford D., and Q. Zhang. 2016. How to save a life: Could real-time sensor data have saved mrs elle? In *Proceedings of the 2016 CHI conference extended abstracts on human factors in computing systems*, ACM, 910–920.
7. Brain, S. 2012. Walmart company statistic. http://www.statisticbrain.com/wal-mart-company-statistics/. Accessed 30 Oct 2016.
8. Brush A., B. Lee, R. Mahajan, S. Agarwal, S. Saroiu, and C. Dixon. 2011. Home automation in the wild: Challenges and opportunities. In *Proceedings of the SIGCHI conference on human factors in computing systems*, ACM, 2115–2124.
9. Burrows, A., R. Gooberman-Hill, and D. Coyle. 2015. Empirically derived user attributes for the design of home healthcare technologies. *Personal and Ubiquitous Computing* 19(8): 1233–1245.
10. Chae S., Y. Yang, J. Byun, and T.D. Han. 2016. Personal smart space: Iot based user recognition and device control. In *2016 IEEE Tenth International conference on semantic computing (ICSC)*, IEEE, 181–182.
11. Cisco. 2013. Internet of everything. http://ioeassessment.cisco.com/. Accessed 19 Oct 2016.
12. Civitarese, G., S. Belfiore, C. Bettini. 2016. Let the objects tell what you are doing. In *Proceedings of the 2016 ACM international joint conference on pervasive and ubiquitous computing: Adjunct*, ACM, 773–782.
13. Cook, D.J. 2012. How smart is your home? *Science* 335(6076): 1579–1581.
14. Cook, D.J., and N.C. Krishnan. (2015). *Activity learning: Discovering, recognizing, and predicting human behavior from sensor data*. New York: Wiley.
15. Cook, D.J., G.M. Youngblood, E.O. Heierman III, K. Gopalratnam, S. Rao, A. Litvin, and F. Khawaja. 2003. Mavhome: An agent-based smart home. *PerCom* 3: 521–524.
16. Das M., G. De Francisci Morales, A. Gionis, and I. Weber. 2013. Learning to question: Leveraging user preferences for shopping advice. In *Proceedings of the 19th ACM SIGKDD international conference on knowledge discovery and data mining*, ACM, 203–211.
17. Dawadi P., D.J. Cook, and M. Schmitter-Edgecombe. 2014. Smart home-based longitudinal functional assessment. In *Proceedings of the 2014 ACM international joint conference on pervasive and ubiquitous computing: Adjunct publication*, ACM, 1217–1224.
18. Despouys R., R. Sharrock, and I. Demeure. 2014. Sensemaking in the autonomic smart-home. In *Proceedings of the 2014 ACM international joint conference on pervasive and ubiquitous computing: Adjunct publication*, ACM, 887–894.
19. Ding, D., R.A. Cooper, P.F. Pasquina, and L. Fici-Pasquina. 2011. Sensor technology for smart homes. *Maturitas* 69(2): 131–136.
20. Fogli D., R. Lanzilotti, A. Piccinno, and P. Tosi. 2016. Ami@ Home: A game-based collaborative system for smart home configuration. In *Proceedings of the international working conference on advanced visual interfaces*, ACM, 308–309.

21. Garnier-Moiroux D., F. Silveira, and A. Sheth. 2013. Towards user identification in the home from appliance usage patterns. In *Proceedings of the 2013 ACM conference on pervasive and ubiquitous computing adjunct publication*, ACM, 861–868.
22. Gartner. 2014. Gartner says a typical family home could contain more than 500 smart devices by 2022. http://www.gartner.com/newsroom/id/2839717. Accessed 30 Oct 2016.
23. Greenhalgh, T., J. Wherton, P. Sugarhood, S. Hinder, R. Procter, and R. Stones. 2013. What matters to older people with assisted living needs? a phenomenological analysis of the use and non-use of telehealth and telecare. *Social Science and Medicine* 93: 86–94.
24. Guo, X., E.C. Chan, C. Liu, K. Wu, S. Liu, and L.M. Ni. 2014. Shopprofiler: Profiling shops with crowdsourcing data. In *IEEE INFOCOM 2014-IEEE conference on computer communications*, IEEE, 1240–1248.
25. Huang, Y.C., K.Y. Wu, and Y.T. Liu. 2013. Future home design: An emotional communication channel approach to smart space. *Personal and Ubiquitous Computing* 17(6): 1281–1293.
26. IEEE. 2016. Iot home of the future. http://transmitter.ieee.org/iot/. Accessed 17 Oct 2016.
27. Karamshuk D., A. Noulas, S. Scellato, V. Nicosia, and C. Mascolo. 2013. Geo-spotting: Mining online location-based services for optimal retail store placement. In *Proceedings of the 19th ACM SIGKDD international conference on Knowledge discovery and data mining*, ACM, 793–801.
28. Kato, T. 2011. User modeling through unconscious interaction with smart shop. In *International conference on universal access in human-computer interaction*, 61–68, Springer.
29. Kientz, J.A., S.N. Patel, B. Jones, E. Price, E.D. Mynatt, G.D. Abowd. 2008. The georgia tech aware home. In *CHI'08 extended abstracts on human factors in computing systems*, ACM, 3675–3680
30. Kleinberger, T., M. Becker, E. Ras, A. Holzinger, P. Müller. 2007. Ambient intelligence in assisted living: enable elderly people to handle future interfaces. In *International conference on universal access in human-computer interaction*, 103–112, Springer.
31. Kubitza, T., A. Voit, D. Weber and A. Schmidt. 2016. An iot infrastructure for ubiquitous notifications in intelligent living environments. In *Proceedings of the 2016 ACM international joint conference on pervasive and ubiquitous computing: Adjunct*, ACM, 1536–1541.
32. B.M. Landry, and K. Dempski. 2012. Weshop: Using social data as context in the retail experience. In *Proceedings of the 2012 ACM conference on ubiquitous computing*, ACM, 663–664
33. Lapointe, J., B. Bouchard, J. Bouchard, A. Potvin, and A. Bouzouane. 2012. Smart homes for people with alzheimer's disease: Adapting prompting strategies to the patient's cognitive profile. In *Proceedings of the 5th international conference on pervasive technologies related to assistive environments*, ACM, 30.
34. Lee, S., C. Min, C. Yoo, and J. Song. 2013. Understanding customer malling behavior in an urban shopping mall using smartphones. In *Proceedings of the 2013 ACM conference on pervasive and ubiquitous computing adjunct publication*, ACM, 901–910
35. Leech, J.A., W.C. Nelson, R.T. Burnett, S. Aaron, and M.E. Raizenne. 2002. It's about time: A comparison of canadian and american time-activity patterns. *Journal of exposure analysis and environmental epidemiology* 12: 6.
36. Liu, T., L. Yang, X.Y. Li, H. Huang, and Y. Liu. 2015. Tagbooth: Deep shopping data acquisition powered by rfid tags. In *2015 IEEE conference on computer communications (INFOCOM)*, IEEE, 1670–1678.
37. Longo, S., E. Kovacs, J. Franke, and M. Martin. 2013. Enriching shopping experiences with pervasive displays and smart things. In *Proceedings of the 2013 ACM conference on pervasive and ubiquitous computing adjunct publication*, ACM, 991–998.
38. Mennicken, S., and E.M. Huang. 2012. Hacking the natural habitat: An in-the-wild study of smart homes, their development, and the people who live in them. In *International conference on pervasive computing*, 143–160, Springer.
39. Mennicken, S., J. Vermeulen, and E.M. Huang. 2014. From today's augmented houses to tomorrow's smart homes: New directions for home automation research. In: *Proceedings of the 2014 ACM international joint conference on pervasive and ubiquitous computing*, ACM, 105–115

40. Miorandi, D., S. Sicari, F. De Pellegrini, and I. Chlamtac. 2012. Internet of things: Vision, applications and research challenges. *Ad Hoc Networks* 10(7): 1497–1516.
41. Network I. 2015. 2015 state of the smart home report. https://www.icontrol.com/wp-content/uploads/2015/06/Smart_Home_Report_2015.pdf. Accessed 18 Oct 2016.
42. Park, S.H., S.H. Won, J.B. Lee, and S.W. Kim. 2003. Smart home-digitally engineered domestic life. *Personal and Ubiquitous Computing* 7(3–4): 189–196.
43. PwC. 2016. London to be transformed from city of home-owners to city of home-renters in a generation. http://pwc.blogs.com/press_room/2016/02/london-to-be-transformed-from-city-of-home-owners-to-city-of-home-renters-in-a-generation.html. Accessed 17 Oct 2016.
44. Radhakrishnan, M., S. Eswaran, S. Sen, V. Subbaraju, A. Misra, and R.K. Balan. 2016. Demo: Smartwatch based shopping gesture recognition. In *Proceedings of the 14th annual international conference on mobile systems, applications, and services companion*, ACM, 115–115.
45. Rai H.G., K. Jonna, and P.R. Krishna. 2011. Video analytics solution for tracking customer locations in retail shopping malls. In *Proceedings of the 17th ACM SIGKDD international conference on knowledge discovery and data mining*, ACM, 773–776.
46. Rashidi, P., and A. Mihailidis. 2013. A survey on ambient-assisted living tools for older adults. *IEEE Journal of Biomedical and Health Informatics* 17(3): 579–590.
47. Runyan, R.C., I.M. Foster, K. Green Atkins, and Y.K. Kim. 2012. Smart shopping: Conceptualization and measurement. *International Journal of Retail and Distribution Management* 40(5): 360–375.
48. Stankovic, J.A. 2014. Research directions for the internet of things. *IEEE Internet of Things Journal* 1(1): 3–9.
49. Tapia E.M., S.S. Intille, and K. Larson. 2004. Activity recognition in the home using simple and ubiquitous sensors. In *International conference on pervasive computing*, 158–175, Springer.
50. Taylor, A.S., R. Harper, L. Swan, S. Izadi, A. Sellen, and M. Perry. 2007. Homes that make us smart. *Personal and Ubiquitous Computing* 11(5): 383–393.
51. On the Internet of Things IERC. 2015. Iot semantic interoperability: Research challenges, best practices, recommendations and next steps. Tech. rep. Accessed 19 Oct 2016.
52. Vianello, A., Y. Florack, A. Bellucci, and G. Jacucci. 2016. T4tags 2.0: A tangible system for supporting users' needs in the domestic environment. In *Proceedings of the TEI'16: Tenth international conference on tangible, embedded, and embodied interaction*, ACM, 38–43.
53. Wang, P., J. Guo, and Y. Lan. 2014. Modeling retail transaction data for personalized shopping recommendation. In *Proceedings of the 23rd ACM international conference on conference on information and knowledge management*, ACM, 1979–1982.
54. Wilson, C., T. Hargreaves, and R. Hauxwell-Baldwin. 2015. Smart homes and their users: A systematic analysis and key challenges. *Personal and Ubiquitous Computing* 19(2): 463–476.
55. Yang R., M.W. Newman. 2013. Learning from a learning thermostat: Lessons for intelligent systems for the home. In *Proceedings of the 2013 ACM international joint conference on Pervasive and ubiquitous computing*, ACM, 93–102.
56. Yuan B., M. Orlowska, and S. Sadiq. 2006. Real-time acquisition of buyer behaviour data—the smart shop floor scenario. In *International workshop on business intelligence for the real-time enterprise*, 106–117, Springer.
57. Zeng, Y., P.H. Pathak, and P. Mohapatra. 2015. Analyzing shopper's behavior through wifi signals. In *Proceedings of the 2nd workshop on workshop on physical analytics*, ACM, 13–18.

SMART-FI: Exploiting Open IoT Data from Smart Cities in the Future Internet Society

Stefan Nastic, Javier Cubo, Malena Donato, Schahram Dustdar, Örjan Guthu. Mats Jonsson, Ömer Özdemir, Ernesto Pimentel and M. Serdar Yümlü

Abstract Smart Cities of the future have a potential to serve as a holistic platform for generating values from the abundance of currently untapped human, societal and ICT capital. Currently, Smart Cities are ever-stronger facing numerous challenges and a stringent need to optimize their urban processes, infrastructure and facilities, such as urban transportation and energy management. Unfortunately, at the moment, small portion of urban data is being exploited for gaining better insights and optimizing Smart City processes. In this chapter, we introduce a novel Smart City platform being developed in the context of SMART-FI project. The SMART-FI platform aims to facilitate analyzing, deploying, managing and interoperating Smart City data analytics services. Firstly, SMART-FI strives to enable collecting the data from a variety

S. Nastic · S. Dustdar
Distributed Systems Group, Tu Wien, Austria
e-mail: nastic@dsg.tuwien.ac.at

S. Dustdar
e-mail: dustdar@dsg.tuwien.ac.at

J. Cubo (✉) · E. Pimentel
Dpto. Lenguajes Y Ciencias de la Computación, Universidad de Málaga, Málaga, Spain
e-mail: cubo@lcc.uma.es

E. Pimentel
e-mail: ernesto@lcc.uma.es

M. Donato · Ö. Özdemir
Atos Reserach and Innovation Unit in Atos Spain Group, Madrid, Spain
e-mail: malena.donato@atos.net

Ö. Özdemir
e-mail: omer.ozdemir@atos.net

Ö.G. Mats Jonsson
NetPort Science Park, Karlshamn, Sweden
e-mail: mats.jonsson@netport.se

M. Serdar Yümlü
SAMPAS Information & Communication Systems, Istanbul, Turkey
e-mail: serdar.yumlu@sampas.com.tr

B. Di Martino et al. (eds.), *Internet of Everything*, Internet of Things,
https://doi.org/10.1007/978-981-10-5861-5_7

of sources, such as sensors and public data sources. Secondly, the platform provides mechanisms for homogenizing the data coming from various networks and protocols. Finally, it provides facilities to develop, deploy and orchestrate novel, added-value Smart City data analytic services. To demonstrate the practical feasibility of the proposed solutions and showcase their benefits for the variety of involved stakeholders, SMART-FI will be piloted in three cities: Malaga (Spain), Karlshamn (Sweden), and Malatya (Turkey).

1 Introduction

A concept long developed, the idea of a connected city [12, 23]—in which our daily lives are augmented by automated technology—is fast becoming a reality. Captured within the Internet of Things umbrella, Smart Cities expose large amounts of urban smart devices to interconnect them and exploit their capabilities, and use digital technologies to optimize the efficiency of municipal processes in order to reduce costs and, most importantly, improve the wellbeing of their citizens [25, 31]. While there is no a single accepted definition, the common contemporary understanding of a Smart City assumes a coherent urban development strategy to manage various city's infrastructural assets and municipal services enhancing citizen's lives and making cities better and sustainable places. These complex and crucial challenges are not only at the technological level, but also addresses sociological and political aspects. Therefore, there is a necessity of having an infrastructure to manage the varying density of data in devices and services using them. Future Internet applications will manage Big Data to include people interacting with the physical world providing facilities for smart cities. In this sense, the importance of ICT role in the Smart City vision is obvious and it ranges from collecting and analyzing data, predicting and optimizing business processes, as well as facilitating communication between different city services and automated management of infrastructure. Unfortunately, at the moment, small portion of the available city data is being exploited for gaining better insights and optimizing Smart City processes. This is mainly due to current challenges that hinder city-scale data analytics solutions including: (i) lack of common data formats, (ii) lack of scalable infrastructure to manage volume, variety, velocity, and veracity of data in urban smart devices, (iii) insufficient support for developers to use, deploy and interoperate services, and (iv) lack of holistic ecosystem that enables the citizens to be agnostic of the technologies behind the solutions, and at the same time allows them to take advantage of such solutions.

Smart Cities have the potential to serve as a platform unleashing generation of values which rise from the abundance of currently untapped human, societal and ICT capital reaching beyond physical city boundaries. In this chapter, we introduce a novel Smart City platform that is being developed in the context of SMART-FI.[1] The SMART-FI platform will facilitate analyzing, deploying, managing and

[1]The SMART-FI project: http://smart-fi.eu/ project.

interoperating Smart City services by exploiting aggregated open data from smart cities. SMART-FI uses smart cities data to provide services on top of FIWARE infrastructure. The main aim of the proposed platform is to develop a set of facilities to provide: (i) methodologies to homogenize heterogeneous Smart City and Open Data, (ii) models and tools for developing data analytics services that can be used to predict patterns and make recommendations, and (iii) mechanisms for data analytics services deployment and orchestration. The SMART-FI solution is intended for a variety of Smart City stakeholders, such as service developers and their users, all citizens, city representatives and anyone who wishes to benefit from smart cities data.

Remainder of this chapter is organized as follows. In Section 2, we present motivating scenarios based and SMART-FI pilots. Section 3 discusses the related work. In Sect. 4, we introduce the SMART-FI ecosystem for smart cities, discuss its main objectives and relation to FIWARE platform. Section 5 gives an overview of SMART-FI Smart City platform and presents its main components. Section 6 presents the evaluation strategy of SMART- FI solutions. Finally, Sect. 7 concludes the chapter, discusses the expected impact and gives an outlook of the future research.

2 Motivation

To better motivate our work and the stringent need to offer a comprehensive Smart City platform, that enables seamless integration, aggregation and analytics of diverse city data, in the following we present three motivating scenarios derived from SMART-FI's pilot cities.

2.1 Application Scenarios

SMART-FI will test its feasibility on real smart city scenarios. The project has three cities to run its pilots: Malaga, Spain (Urban Transportation), Malatya, Turkey (Smart Society Services) and Karlshamn, Sweden (Urban Energy).

Malaga case study—The use case to be piloted in Spain is based on Malaga city. According to the IDC Smart Cities Index Ranking,[2] Malaga is one of the top five "smartest" city in Spain, with population over 570,000. Malaga has lead the Spanish ranking because is currently an Urban Lab and R&D hub[3]; the city is pioneer in the development of an eco-efficient and sustainable city project based on the optimal integration of renewable energy.

---

[2]https://www.idc.com/.

[3]http://www.urbanlabmalaga.es.

CityGO application is being developed within SMART-FI project in order to offer a smart mobility solution that aims at providing recommendations, such as the best route to get usual destinations within a city, depending on diverse real-time conditions, promoting healthy and environmental care behaviours. Concretely, the CityGO application promotes transportation diversity in the Malaga City through the citizen cooperation using their smartphones and GPS. Further, by utilizing city-wide sensor network and open data from the Malaga City it can get information on urban transportation such as how many parking spaces are available. Based on this data the application calculates the usual daily paths/itineraries, providing also valuable information to the municipality to take decisions about transport regulation. A calculation is further enhanced with daily itineraries and frequency, enabling it to provide predictive information on people's regular transportation habits. The user gets recommendation about the best itinerary based on real time information: such as choosing a bus, rent a bicycle or use the car. It gives that information proactively, reducing the level of user involvement. The CityGO application will help addressing transport problems such as traffic and parking congestion, and pollution in cities, and the stress associated to these. The usability of an application with such features supports economic development as could attracts tourists in a given city. Finally, CityGO is expected to increases city's livability as it can contribute to developing healthier and less stressful living environment.

Malatya case study—Malatya[4] is a metropolitan area in Turkey with its population over 750,000. In order to provide better connections and increase cooperation between the city authorities and its residents, Malatya Metropolitan Municipality started its Smart City Initiative that promotes the use of government facilities using the largest possible number of smart applications and implementing projects on four main domain: society, mobility, governance and environment. With the ultimate goal of being connected, providing optimum resource management and an immense opportunity for providers of smart solutions, Malatya have been awarded by The World e-Governments Organization of Cities and Local Governments (WeGO) with its ICT focused smart city projects including MAKBIS,[5] Malatya Intelligent City Automation System.

The aim of the SMART-FI use case in Malatya, piloted through MalatyaInsight application, is to have a better understanding of citizens' needs and priorities while providing high quality, transparent and responsive services for citizens and an opportunity for developers to develop innovative applications. This application is geared towards the "Smart Society" and it is based on governance services, integration and participation processes. It will facilitate participation, inputs, and ideas from a wide range of stakeholders in the city. Based on this, the MalatyaInsight application aims to provide an open data portal based on SMART-FI and a mobile interface that provides accurate, normalized and real information about several governance services and investment projects in Malatya. It also provides accurate business industry data for citizens, enabling them to participate by providing comments, ratings, demands

[4]http://www.malatya.bel.tr.

[5]http://www.malatyagelisiyor.com/proje/16/33/makbis.aspx.

and complaints related to the business, as well as getting personalized recommendations based on the time, location, their historical records and interested keywords for both citizens and tourists in Malatya. The MalatyInsight will be the first open data portal for Malatya, promoting and enabling a development of a society with better living conditions.

Karlshamn case study—In Karlshamn we have two independent use cases that address public transport and smart buildings, both focused on urban energy sector. Karlshamn is a small municipality in the south east of Sweden with some 31,000 inhabitants. The activities will specifically address the needs confronted by smaller municipalities.

The first application to be developed is related to public transportation. The application, BlekingePublicTransit, will use data collected from buses in real-time to collect, analyse and present for two categories of users, the traffic planner and private persons using the transportation services. Services to enable more efficient travelling and use of resources for both parts will be tested and presented. Some examples of services are "when is the next bus coming" for the traveller or "is there sufficient space between the buses" to create an efficient flow for the traffic planner. In all, this kind of services will provide incentives for the traveller as well as the traffic operator to move in the direction of more efficient energy consumption.

The second application, SmartBuilding, is related to smart buildings where a new technology to collect data from buildings, such as power consumption, air quality and person movement will be used. The data will be presented both for the building manager and the people using the office space in order to make better data-driven decisions and provide inputs for more efficient use of energy and change of user behavior. Some examples of services are the possibility to measure energy consumption in each electricity outlet to visualise for the user or adjusting the room climate to meet personal preferences as people are entering the room. The implementation of this use case evaluates this approach in a limited scale regarding the energy consumption. The principles though are applicable in a much larger scale. With equivalent systems and components, most buildings in a large city could be optimized regarding all the electrical functions, such as e.g. heating, ventilation, lighting and security systems. For instance, by unifying the data from lighting control systems, thermostats and other sensors the application will enable automatically adjusting building settings according to real-time usage patterns, leading to energy savings, improved air quality, and an increase in overall efficiency.

Based on the described case studies, at the moment some of the biggest challenges for municipalities is formulating a coherent responses to exploit the opportunities of the available urban data, thus preventing harnessing the potential benefits. This is mainly due to heterogeneity of data sources and formats, a lack of data analytics capabilities and suitable tools for intuitive development of intelligent Smart City applications.

3 Related Work

In this section, we present related work containing research efforts and projects, at different levels, as regards the main functionality provided by SMART-FI: (i) Data normalization, (ii) Data analytics, and (iii) Service orchestration.

As stated in the Digital Agenda for Europe [21], generating value at different stages of the data value chain will be at the center of the future knowledge economy. Good use of data can bring opportunities also to more traditional sectors such as transport, resources management, health, agriculture or manufacturing. However, the process of opening data is a necessary but complex task for cities. Norms such as UNE 178301:2015 (Smart Cities. Open Data) [2] are proposing ways of measuring the maturity of a city opening their data. Even though, these norms can be used to guide the process of opening data, they do not propose methodologies to reach maximum maturity levels. Instead they are usually mostly focusing on providing semantics to sensor data methods [34].

Linked Open Data (LOD) has gained significant momentum over the past years as a best practice of promoting the sharing and publication of structured data on the semantic Web [11]. In the last few years, several works tried to focus on Semantic Web technologies for extending and integrating smart city data systems. Semantic Web technologies enable to put into practice the Open Government Data (OGD) principles of transparency, participation and collaboration, in the purpose to integrate citizen within the smart city paradigms [1, 7, 10]. Public authorities and local governments started to share their open data sets. Km4Cityn [8] provided the process adopted to produce the ontology and the big data architecture for the smart city knowledge base and showed the mechanisms of data verification, reconciliation and validation. Consoli et al. [17] understanding the importance of a good data model in smart city applications, provided a platform that exhibits a consistent, minimal and comprehensive semantic data model for the city based on the Linked Open Data paradigm. They used W3C standards and good practices of ontology design in order to solve data management and representation issues. However, most of these large volumes of data sets are not semantically interoperable, thus making it practically challenging to generate a cocherent knowledge base for smart cities.

Regarding opening data silos and architectures for public authorities and semantic to heterogeneous open data sets and USDL languages, EU have funded several initiative projects like SOA4ALL,[6] OPEN-DAI[7] and OPTIMIS.[8] The SOA4All project provides a comprehensive global service delivery platform that integrates complementary and revolutionary technical advances into a coherent and domain independent service delivery platform including SOA, Web 2.0 and Semantic Web Technologies. Open-DAI (Opening Data Architectures and Infrastructures of European Public Administrations) aims to make data and platforms available for digital public services on cloud computing infrastructures. Open-DAI represents a new model in

---

[6]http://cordis.europa.eu/fp7/ict/ssai/docs/fp7call1achievements/soa4all.pdf.

[7]http://open-dai.eu/.

[8]http://optimis-project.eu/project.

both new PAs services implementation and cloud model deployment. OPTIMIS is aimed at enabling organizations to automatically externalize services and applications to trustworthy and auditable cloud providers in the hybrid model. OPTIMIS delivered a specification and toolkit that is used to construct next-generation cloud architectures also for cities.

With respect to the data analytics, various underlying data processing frameworks, especially for streaming and batch analytics exist. Among the seminal papers, Agrawal et al. [3] and Cuzzocrea et al. [19] explored the possibility of running database management systems (DBMS) in a Cloud environment, concluding that there is not a one-size-fits-all solution, due to the trade-off between scalability and query expressiveness. The availability of huge amount of daily produced Smart City data fosters the exploration of new data analytics approaches which should simplify the ingestion, transformation, and consumption of data by possibly neglecting (or hiding) the complexity of managing a scalable and distributed processing system. Supported by concrete case studies, Hashem et al. [22] nicely highlight the tied relationship that exists between Cloud computing and Big Data. The former provides the underlying engine that enables several classes of distributed data-processing platforms (e.g., batch processing, stream processing), whereas the latter might utilize distributed and fault-tolerant storage technologies based on Cloud resources in order to simplify the management and processing of data. Moreover, some research efforts envision and conceive a conceptual architecture that tightly combines the requirement of data analytics and the potentialities of Cloud computing. It results a model named *Cloud-based Analytics as a Service* or *Data Analytics as a Service*. For example, Domenico Talia [32] discussed the complexity and variety of data types and processing power to perform analysis on large datasets. Therefore, he proposed three Cloud-based service models that support their execution: *data analytics software as a service*, where an analytic application or task is offered as a service, *data analytics platform as a service*, where analytic suites or frameworks are offered hiding the cloud infrastructure, and *data analytics infrastructure as a service*, where virtualized resources enable the storage and processing of Big Data. This idea is exploited also in other research papers, e.g., [20, 36], among which the one by Zulkernine et al. [36] presents a conceptual architecture for Cloud-based Analytics-aaS, which however includes only a preliminary implementation, lacking the details of how to process the massive dataset. Other research contributions focus more on the scalable execution of user-defined functions (UDF) among a large and distributed set of computing nodes, which can be acquired or released as needed, encompassing the on-demand resource principle of Cloud computing. Nowadays, two opposite approaches are commonly adopted: batch processing and stream processing [27]. The former stores all the data, usually on a distributed file system, and then operates on them on the basis of different programming models, among which the well-known MapReduce. The latter processes all the data on-the-fly, i.e., without storing them, so it can produce results in a near real-time fashion. A plethora of frameworks is available to process the data following one or the other approach: examples of batch processing frameworks are Apache

Hadoop[9] (an open-source implementation of MapReduce) and Apache Tez.[10] Examples of stream processing frameworks are Apache Storm [33], Apache Flink,[11] IBM Infosphere [9], and Amazon Kinesis.[12] Finally, there have been several initiatives in European Union under 7th Framework Programme and Horizon 2020. The SUPERSEDE[13] project proposes a feedback-driven approach to the life cycle management of software services and applications, with the ultimate purpose of improving users' quality of experience. Decisions on software evolution and runtime adaptation will be made upon analysis of end-user feedback and large amount of data monitored from the context. An integrated platform will articulate the methods and tools produced in the project. MARKOS[14] project uses data analytics techniques for the analysis of software sources, and the support to consume these data, migrated to a RDF/S repository and accessing through SPARQL Endpoints. All of these solutions could be used to execute data analytics processes. However, there is a lack of tools to generate/accelerate elastic data analytics services that utilize these frameworks to handle large-scale data to offer new analytics under micro services models. Most of the time, the developer has to write all analytics functions, service interfaces and complex configurations for elasticity.

With respect to service orchestration is the integration of more than one service to work together. Service composition will create apps generated in form of orchestrators or adaptors with new functionalities to be remotely accessed (SaaS or Mashups), which could be deployed in infrastructures (Clouds). The incremental construction of Future Internet apps through the adaptation and orchestration of reusable services specified with interfaces is an error-prone task. It is possible to assist developers with automatic procedures and tools supplied by model-based software adaptation. But, in most cases it is impossible to modify the realization of services to adapt them, since their internal implementation cannot be inspected/modified. Due to the services' black-box nature, they must be equipped with external interfaces with information about their functionality. Interfaces of services of a system do not always fit and some features of these services may change at run-time [13], so they require an adaptation [16, 29] to avoid mismatching, detected with monitoring [1]. It is required to study the compatibility of services [5, 35], considering their evolution [4]. Current solutions for service integration and interoperability are divided between restrictive [6, 14, 28, 30] and generative [15, 18, 26]. The former approaches are focused on restricting the system to a desired subset of interaction traces. Generative ones create new possibilities (traces), which were not originally available in the system. Instead of restricting to avoid failed interactions, they support new communications so as to make every possible interaction successful. Furthermore, some of the recent funded EU projects concentrate on providing techniques for the adaptation and integration

[9]http://hadoop.apache.org/.

[10]https://tez.apache.org/.

[11]https://flink.apache.org/.

[12]https://aws.amazon.com/kinesis/.

[13]https://www.supersede.eu/.

[14]http://www.markosproject.eu.

issues including modeling and design techniques such as MODACLOUDS[15] and ARTIST.[16] Another approach taken by projects like SUPERSEDE and SOA4ALL[17] approaches in a different way by providing techniques of composition for orchestration and the integration through an Enterprise Service Bus. Among recent research studies CLOUDSOCKET[18] and IOT.est[19] provides a third of line approach for the orchestration, i.e. composition, of business services based on re-usable IoT service components and the abstraction of the heterogeneity of underlying technologies to ensure interoperability by providing workflow engines and integrated frameworks respectively. SMART-FI approach is building on the existing solutions to produce methods for deploying and integrating existing or new services and applications, for producing more advanced applications with the composition and orchestration of simpler ones (mashups of services). As a result a marketplace will be generated by the project and third-party applications giving value to the public data.

4 SMART-FI Ecosystem for Smart Cities

4.1 Smart City Ecosystem Requirements

Recently, the EU has been promoting the Smart City digital single market[20] to have digital technologies to serve citizens with the goal to have better public services, better use of resources and less impact on the environment. SMART-FI is expected to play a major role in small but embryonic smart cities by implement the three use cases (cf. Sect. 2). Although the three cities, piloting the SMART-FI solutions, are not so big considering the number of inhabitants, they are facing many economic, ecological and societal challenges that with the help of technology could be overcome. In the following, we present the main smart city ecosystem requirements derived from the aforementioned use cases and discuss them in the broader context of SMART-FI ecosystem.

Given the scope and diversity of the involved stakeholders and the potential impact of future Smart Cities, it is clear that a comprehensive ecosystem, spanning beyond mere technical solutions, e.g., a platform is required. Such ecosystem should be complete enough to serve people's needs offering a marketplace to use data from cities (i.e. transportation, energy, smart society data, etc.), and use data analytics services so that the third party could reuse these. Additionally, a Smart City ecosystem should contain a structured roadmap and a set of guidelines on how to realize, implement

[15] http://www.modaclouds.eu/.

[16] http://www.artist-project.eu/.

[17] http://projects.kmi.open.ac.uk/soa4all/.

[18] https://www.cloudsocket.eu/.

[19] http://cordis.europa.eu/project/rcn/99934_en.html.

[20] https://ec.europa.eu/digital-single-market/en/smart-cities.

and deploy these services in practice. The system should be intuitive enough and unobtrusive to use as this is for people who are familiar with solutions but yet they deserve simple and user friendly apps.

On the other hand, a suitable Smart City platform should be capable to satisfy a set of technical requirements. Based on the presented use cases in Sect. 2, we have derived a set of such requirements that include: (i) A data normalization solutions that allow integrating the variety of data coming from various sources (e.g., sensors, public services and open data), in different data formats and via different protocols. (ii) A structured support for Smart City application developers, in order to lower the barrier of application development and facilitate development of data analytics services that serve as reusable building blocks for complex Smart City applications. (iii) A generic, extensible and scalable platform infrastructure, that allows easier handling of volume, variety, velocity, and veracity of urban data.

Therefore, it seems evident that interacting with the virtual and physical environment at different levels, is a way to describe the SMART-FI ecosystem considering the smart city areas where the platform aims to provide value. There is an obvious need to have an infrastructure highly scalable to manage the varying density of data in urban smart devices and services, including people interacting. The current data sources heterogeneity or even the lack of common data formats prevents the uptake of innovative cross-domain smart city applications and the SMART-FI platform focuses on addressing these issues.

4.2 *Overview of SMART-FI Approach*

Smart Cities are considered open innovation ecosystems, and the SMART-FI ecosystem aims to facilitate interaction between technology, governing institutions and citizens in order to enable exploiting the opportunities of the Future Internet and smart urban environments. The smart city concept does not limit its range of action to just providing better services. It is tightly correlated with many other aspects associated with the city ecosystem and its stakeholders. Hence, the availability of a platform which enables the support of capabilities beyond service exploitation and optimization is perceived as a unique opportunity for the definitive acceptance of the smart city phenomena. The consideration of issues related to citizen participation, SME involvement and entrepreneurship support are just a few examples of key aspects to be prioritized. For example, today most people use smartphones to interact with the world they live in to plan, schedule and organize their lives; all this information is available right within the palm of a hand. Citizens would like to have accurate and real-time personalized solutions from information regarding immediate or even near-future traffic, mobility patterns, participating through comments in city investment projects, or rating a specific touristic place, or other services relevant for their daily life.

The SMART-FI approach is expected to help deploying and interconnecting services setting up the right technology and using real open data from diverse sources, mainly from public administrations, but also from other third-party services or devices. The aim is to provide services on top of FIWARE, an standard open IoT platform recognized at EU level, that facilitates the development of smart applications and with an environment where cities can publish their open data. SMART-FI main goal is to create a platform and a set of facilities to deploy and interoperate services by exploiting aggregated open data from smart cities. The project will provide methodologies to homogenize heterogeneous open data and data services, perform analysis and aggregation of data analytics services to predict patterns and make recommendations, as well as to facilitate services deployment and composition by orchestrating different services and applications. Additionally, the SMART-FI solution aims at supporting local authorities, public transport operators and other organisations to optimize the services provided supporting the concept of a smart city. Thus, opening a path to allow opportunities for third parties to offer services.

This is the reason why in smart city ecosystem, a platform needs to support interaction between humans and devices in three consecutive activities: collecting, communicating and using information. First, the approach in SMART-FI is to collect information through city sensors, mobile devices or directly from different services to capture data such as location, routes, parking availability, or even temperature, sound, location etc. Secondly, it communicates that data using different networks. And finally, it interprets data to understand what has happened and what is likely to happen next, making predictions and recommendations for citizens.

4.3 FIWARE Platform and SMART-FI Ecosystem

As previously mentioned, SMART-FI will produce a set of facilities aligned to FIWARE platform. As an open source platform, supported by companies, universities and research institutions, the FIWARE platform will play a key role in the cities of the future.[21] Its massive adoption may help to speed up the replication of key components for setting up and consolidating the smart city ecosystem. The FIWARE cloud and software platform is a good catalyst for an open ecosystem of entrepreneurs aiming at developing state-of-the-art data-driven applications. This ecosystem is formed by application developers, technology and infrastructure providers and entities who aim to leverage the impact of developing new applications based on the data they produce and publish. In this context cities will play a unique role, especially those implementing Smart City strategy, which open up their data to facilitate the creation of applications built by developers that form part of this ecosystem. FIWARE enables quick and easy application development because they make use of prefabricated components known as generic enablers in its cloud, sharing their own data as well as accessing open data from cities.

[21] https://www.fiware.org/tag/smart-cities/.

In SMART-FI platform will uses several FIWARE generic enablers, such as CKAN, COSMOS, IDM, Cygnus and Orion Context Broker. For example, CKAN is intended for data publishers, e.g., national and regional governments, companies and organizations that want to make their data open and available. IDM covers a number of aspects involving users' access to networks, services and applications, including secure and private authentication from users to devices, networks and services, authorization and trust management, user profile management, privacy-preserving disposition of personal data, Single Sign-On (SSO) to service domains and Identity Federation towards applications. IDAS enable connecting IoT devices/gateways to FIWARE-based solutions, by translating IoT-specific protocols into a standardized NGSI context information protocol. Finally, Wirecloud GE builds on cutting-edge end-user development, RIA and semantic technologies in order to offer a next-generation end-user centred web application mashup platform aimed at leveraging the long tail of the Internet of Services.

5 Smart-FI Platform

In this section, we present SMART-FI platform architecture overview. The SMART-FI platform is the central facility of the SMART-FI ecosystem that serves as one of the main building blocks and a cornerstone for developing a sustainable SMART-FI ecosystem. It enables developing, managing and interoperating Smart City data analytics services, in order to facilitate exploiting Smart City open data and optimizing various city sectors, such as transportation, governance services and urban energy. The main objective of the SMART-FI platform is to allow horizontal integration of open city data and data analytics services by providing a set of generic component and mechanism that will enable development of higher-level, added-value Smart city applications and services.

As discussed in Sect. 4.3, the SMART-FI platform is based on the FIWARE and it utilizes its several components. However, it goes one step beyond by developing generic components and mechanisms based on microservices technologies, in such a way that other Smart City platforms could seamlessly adapt and incorporate SMART-FI components to suit their needs. Also, the supported facilities could be extended in a future, in a simple way, for instance, in order to include a more complex service for data mining, or for service deployment using cloud providers.

Figure 1 shows the SMART-FI platform architecture overview. The SMART-FI platform follows a layered architecture with the main layers including: (i) Data normalization, (ii) Data analytics micro services, and (iii) Services orchestration.

At the physical level, we consider the data sets are coming from different sources, such as public services, devices, and the possible smart physical infrastructure installed in specific cities. FIWARE would be our intermediate piece in order to acquire the data and services. The main FIWARE Generic Enablers and components are depicted in Fig. 1 (they were described in more details in Sect. 4.3).

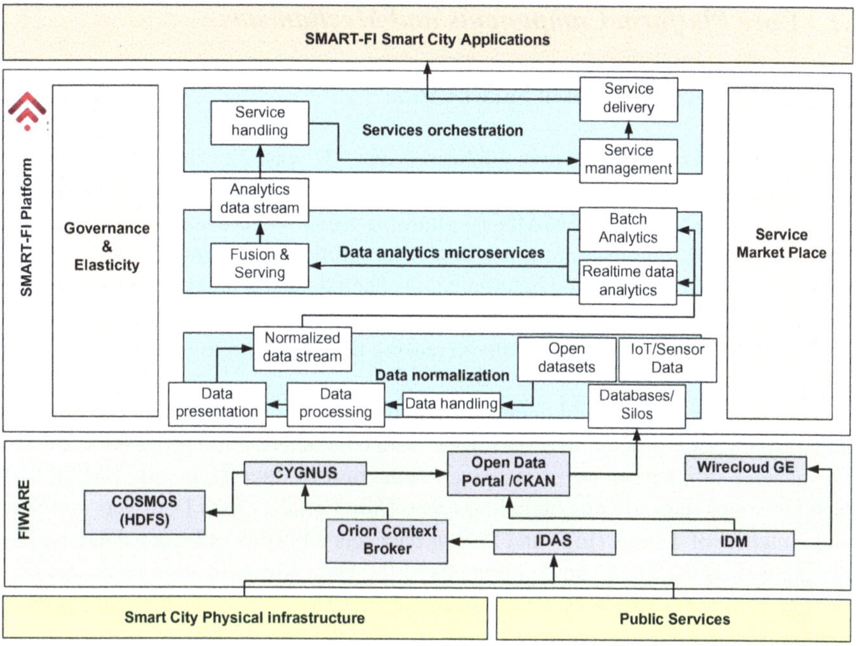

Fig. 1 SMART-FI platform architecture overview

At the SMART-FI platform facilities level, three layers represent the main components, as well as a Governance and Service Market Place. Each facility is capable to perform a set of processes to get its main goal. Here we describe briefly the main flow that will be detailed for each component in Sect. 5.1. Heterogenous data sets will be managed in the Data normalization component. It generates data sets with access rules and a normalized schema, based on urban ontologies, which are stored in semantic data store. Next, these normalized data streams are used by the Data analytics microservices and Service orchestration components. The data analytics services enable the development and management of data analytic services in Smart Cities, providing elastic data analytic services to analyse the aggregated data for predictions and recommendations, as well as developing and managing the called Micro Data Analytic Services (MiDAS). With the Service orchestration facility, mechanisms for deploying and integrating existing or new services and applications will be provided, obtaining more advanced and complex applications (mashups of services) and creating a marketplace that considers the FIWARE Lab and third-party applications.

At the application level, our platform will create SMART-FI services to be used for Smart City Applications.

5.1 Core Platform Components and Mechanisms

5.1.1 Data Normalization in Smart Cities

In order to efficiently create new public services, the vast city data sets should be homogenized and normalized to create urban ontologies where data services will based on. For this purpose, SMART-FI platform uses Linked Data technologies and the Linked USDL language, in order to give structure and semantics to urban environment and smart city related heterogeneous open data sets and data-as-a-services, generating SPARQL Endpoints. This will enable the development of third-party applications taking advantage of data increasing their exploitation and advanced use by citizens.

Figure 2 shows the general component architecture for the Data Characterization and Normalization process in SMART-FI. Data Characterization provides a concise and succinct summarization of the given collection of data. Generally SMART-FI considers cities data streams including Open Data Portals, Open Data Sets and Services, Internet of Things (IoT) and Sensor data sets and silos of legacy databases as data input streams. The main components of the Data Normalization layer include: (i) *Data Handling*, (ii) *Data Processing* and (iii) *Data Presentation* components. Next, we discuss the first two component in more details.

Data Handling layer will receive heterogeneous data sets via the Input Handler and apply pre-processing to produce a list of data sets with access rules and a schema. The data will reside in semantic data stores. During the data processing period, Metadata processing, SPARQL based Query processing and Linked Open and Linked Sensor Data processing techniques will be applied using several different Urban Ontologies to create Urban Environment Ontology and the Data Presentation layer will provide Data Services to both Analytics components and Service Orchestration components.

Data processing layer will homogenize the data using urban ontology definitions and will provide a uniform data format. SMART-FI Data Normalization components will ease the discovery, the standards based normalized data sharing and reduce redundancy by also adding value to build ecosystems around cities data and contents.

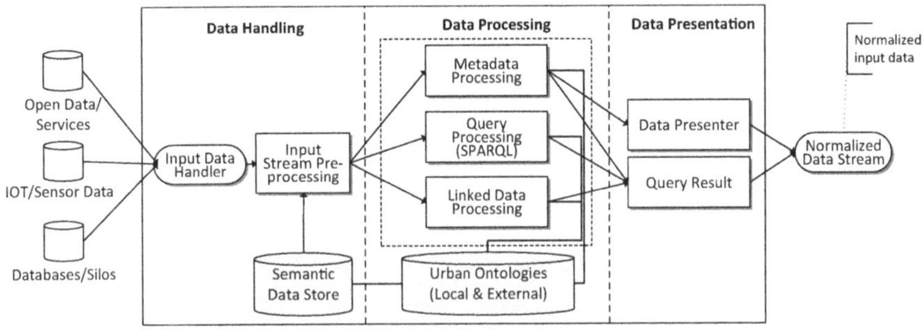

Fig. 2 Data normalization process

This component also includes Linked Data processing capabilities which is one of the best means for publishing and interlinking structured data for access by both humans and machines via the use of the RDF (Resource Description Framework) family of standards for data interchange and SPARQL for query. Here SPARQL-based solutions will be used to facilitate the discovery compared to conventional search mechanisms.

5.1.2 Data Analytics Microservices for Smart Cities

Whereas the data normalization resembles the bloodstream of the SMART-FI platform, the data analytics represents its vital organs. Generally, the main purpose of Smart City data analytics services is enabling transforming the city data into disruptive innovation building blocks for the Smart Cities of the future, based on micro services technologies. To this end, this part of the platform provides models and components that allow for developing and managing value-added data analytic services in Smart Cities. Its purpose is twofold: First, it provides advanced models for programming generic, elastic data analytic services, in order to facilitate analyzing the aggregated data for predictions and recommendations. Second, it implements tooling support for developing and managing such Micro Data Analytic Services (MiDAS).

Figure 3, shows the architecture overview of the micro data analytics services that are capable to support both online and offline Smart City data analytics in a uniform way. The overall data analytics architecture of SMART-FI platform is based on the Lambda architecture [24] It comprises three main layers: *Realtime Data Analytic Layer*, *Batch Analytics Layer* and *Fusion and Serving Layer*. The most important components of the data analytics are the micro data analytics services (depicted as shaded boxes in Fig. 3) include: (i) Batch view function, used to precompute static (slow-changing) partial aggregate views. (ii) Stream transformation function, used to compute realtime window deltas (realtime delta views). (iii) Fusion function, used to combine partial aggregate with realtime delta views and serve the results proactively

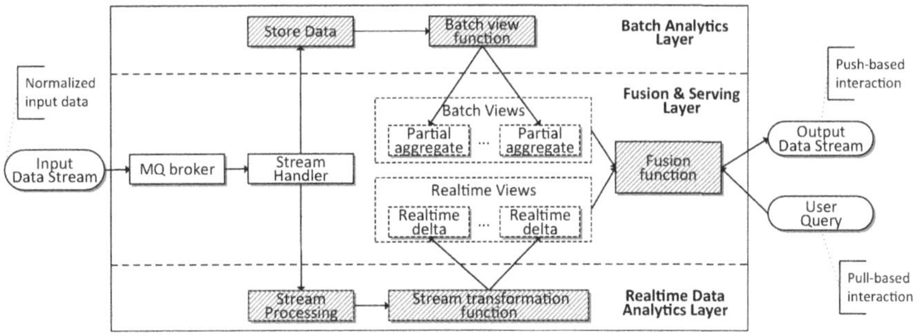

Fig. 3 Micro data analytics services architecture overview

or on-demand, enabling push or pull based interaction. Subsequently, we describe these components in more details, mainly focusing on the *Realtime Data Analytic Layer*. The components depicted as shaded boxes in Fig. 3 represent the computed data views. They are not directly exposed to users and serve as inputs to the Fusion functions. We describe SMART-FI's approach to realizing the Fusion functions in more detail in the next section.

In our platform the we provide a novel model for realtime data analytics, which treats the data streams as first class citizens. In general, there is one-to-one mapping between MiDAS and data streams. A MiDAS is a logical entity identified by an ID (e.g., URI) and its model is characterized by the following main three elements:

- *Stream data*: the sequence of events that constitutes the stream. Every new event is handled by the *Stream Processing* component that triggers the update of the downstream MiDAS, i.e., the streams in a dependent relationship with the current one. The events in a stream can be volatile or temporary stored.
- *Stream Transformation function*: this is a stateless function defined by the user, which transforms the incoming events in new events, according to the contract definition. The transformation function is automatically managed by the execution environment to support elastic scaling, runtime governance and QoS.
- *MiDAS contract*: Generally, the contract defines the type of the stream and encapsulates its most important properties, such as operational mode (i.e., window-based, partition-based mode), side effects and SLAs. Therefore, MiDAS contract can be seen as complex data type in a type system, which is related to the data transformation function.

5.1.3 Services Orchestration in Smart Cities

Since it is not possible to predict all the services and application that will emerge in the Smart Cities of the future, cities need an environment that enables innovation and layering of multiple services (data services, analytics services, etc.) on a common infrastructure. This should also allow the introduction of new elements and re-use of existing resources. To ensure the right service is delivered with the right quality and performance to the right users, a mechanisms enabling careful planning, orchestration and assurance is required. By facilitating orchestration and integration of different public services, several business and Smart City functionalities, coming from different services, are exposed to the end users as a single service endpoint or as a comprehensive Smart City application.

To this end, SMART-FI platform provides methods and tools for deploying and integrating existing or new services and applications, for producing more advanced applications with the composition and orchestration of simpler ones (mashups of services). In order to provide assurance, SMART-FI also aims to create a marketplace that will be generated or enriched considering the Store in FIWARE and third-party applications giving value to the public data.

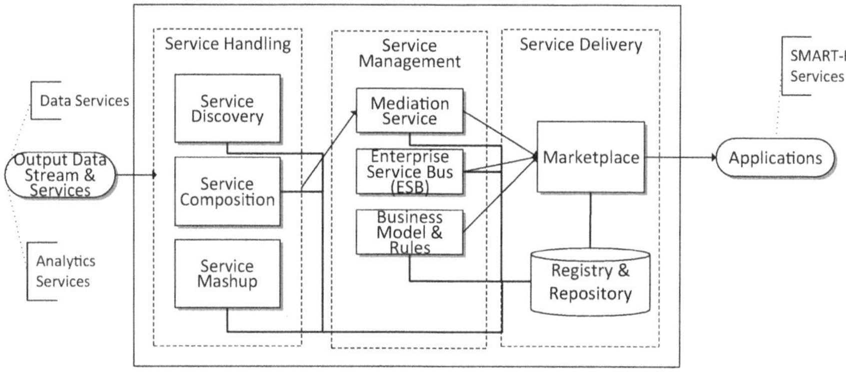

Fig. 4 Service orchestration architecture

For this purpose, SMART-FI platform provides model-based techniques to facilitate the interoperability among applications and methodologies for orchestration and adaptation to create an integrated service development, implementation, deployment and management framework to ensure the governance on the creation and operation of services and their testing procedures. SMART-FI will also support: (i) discovery of added-value combinations of service components and automatically composing and refining these to the needs of the user, (ii) delivery of these composed services at the right time and right place across different platforms and devices in a qualitative and dependable way, and (iii) use of rigorous and lightweight model-based techniques to facilitate interoperability among applications running in an isolated based on orchestration and adaptation methodologies.

Figure 4 shows the basic architecture of service orchestration in SMART-FI platform. It has three different layers when enabling different set of data and analytics services to be provided for the use of smart city based applications. Service Handling is responsible for handling output data stream and services that may come from normalized data services or data analytic services. It comprises different characteristics of Service Discovery, Service Composition and Service Mashup delivery. Service Management is responsible for the mediation of services, business rules and Enterprise Service Bus (ESB) management. The Mediation Service here is a middleware component responsible for providing interoperability among different communication protocols and among different data models. For the effective ServiceDelivery all services all combined in the Registry and Repository. Composed and integrated services are delivered to the application layer through the Marketplace component.

SMART-FI platform offers generative techniques based on integration services to make them pervasive and transparent to the user. Based on intuitive natural-language queries and user profile, mechanisms will infer the requirements and preferences and discover services to provide the desired functionality. Services will be automatically orchestrated on -demand to fulfill functional and QoS requirements. Mashups will be performed by users of the platform using the provided service orchestration and mediation capabilities. Service orchestration and exposure process is going to be achieved with the usage of Enterprise Service Bus (ESB) services. ESB service designers will be used to ease the life of smart city professionals to manage two service endpoints and transform the result into a new data or service. Mediator and sequence based components will used for message mediation and flow construction and holding and transferring the series of mediators respectively.

6 Evaluation Strategy

To evaluate the SMART-FI platform the project will create a *criteria-based evaluation plan*. This evaluation plan will contain the criteria for the evaluation, the variables to be measured, the tools to be used, the mechanisms for doing the evaluation, and the procedures for gathering the evaluation data. The evaluation strategy will not only be used to validate the pilot implementations against their requirements, but it will also be used to validate the overall SMART-FI platform, e.g., in terms of the quality of its components and long-term sustainability. By clarifying individual criterion, their priorities and converting those into measurable values, the evaluation process will guarantee continuously quantifying the progress of the SMART-FI project towards achieving its main objectives (cf. Sect. 4) and long-term sustainability.

All the criteria defined in the evaluation plan will be described in detail and supplemented with lists of questions. This will form the baseline for a criteria-based assessment and a checklist to be used for each delivery related to the end product. The criteria-based assessment will contribute with a measurement of quality in a number of essential areas. These areas are derived from ISO/IEC 9126-1 Software engineering—Product quality[22] and, among others, include usability, sustainability and maintainability. Each of these areas comes with a set of sub-characteristic (e.g. changeability), which further will be divided into attributes. To be able to evaluate a degree to which the quality attributes are met, target values for quality metrics will be derived and specified.

As the SMART-FI platform will contain open data sets, public API's and source code, the evaluation will be performed from users of such components, such as operators and developers. The evaluation process is designed to be continuous and iterative, and all iterations will end with compiling software evaluation reports to be able to constantly improve the platform as long as the SMART-FI project is running. This will be vital in validating whether the SMART-FI platform comply with the various characteristics or show the different qualities that are expected. If the defined characteristics are satisfied and the target values are reached, the SMART-FI platform can be considered to be a general and sustainable solution with good usability and maintainability.

[22] http://www.iso.org/.

7 Conclusion and Future Impact

In this chapter we have introduced the work currently being done in SMART-FI project. Since the project is still at an early stage of development, the main focus was on presenting the vision and general approach of *SMART-FI platform and ecosystem* for Smart Cities of the future. We presented the preliminary architecture, main components and facilities of SMART-FI Smart City platform for data collection, aggregation and analytics. We have discussed how SMART-FI platform is able to analyze open data and enable making personalized recommendations for citizens and public utility operators by facilitating development of data analytics mechanisms. Embracing the smart city paradigm, the platform will bring up opportunities for third parties to offer services. SMART-FI feasibility will be tested on three real smart city scenarios. One of the main objective of SMART-FI platform is to deliver its facilities in a generic and loosely coupled manner, enabling other Smart City platforms or ecosystems to seamlessly use SMART-FI facilities, and to allow the future extension with new facilities that could be beneficial for future smart cities.

SMART-FI aims to make the open data from Smart Cities ready to be used, but at the same time it hides the technical complexities of data integration and analytics. In the near future we expect Smart City stakeholders to reap the benefits of SMART-FI platform by exploiting its technological advancements in data collection, data analytics and service orchestration for Smart Cities. For example, having normalized data coming form a variety of Smart City data sources will enable better vertical and horizontal integration of different Smart City sectors. Further, SMART-FI's support for developing and managing micro data analytics services will enable easier development of generic and reusable data analytics components that can serve as building blocks for complex Smart City prediction and recommendation applications. Finally, the presented platform will provide runtime support for Smart City services orchestration and management of elasticity concerns, thus effectively relieving application developers and operators from much of the burden currently faced when dealing with Smart City applications.

Since we expect that these benefits of the SMART-FI platform will have a significant effect already in a near future, we recognize a need for a long-term roadmap and comprehensive methodology for sustainable development of future Smart Cities. Benefits in the long term are ensured, as SMART-FI aims to increase citizen engagement, citizen participation in decision making process within Smart Cities. Municipalities will provide better public services and all that will foster strengthening the economic power of the city by fostering the development of new and innovative services and products to its citizens. For example, opening of the data generated by cities will foster new relationships between citizens and their government. A solution such as SMART-FI presents an immense opportunity to use published data in a meaningful way resulting in better insights and services for citizens, reaching across different Smart City sectors and even beyond the cities' physical boundaries.

Acknowledgements This work is sponsored by Joint Programming Initiative Urban Europe, ERA-NET, under project No. 5631209. The authors alone are responsible for the content.

References

1. Abid, T., M.R. Laouar, H. Zarzour, and M.T. Khadir. 2016. Smart cities based on web semantic technologies. In *Proceedings of the 2016 ACM International Joint Conference on Pervasive and Ubiquitous Computing: Adjunct*, UbiComp '16, 1303–1308. New York, NY, USA, ACM.
2. AENOR. 2015. UNE 178301:2015. http://bit.ly/2g6LoeN.
3. Agrawal, Divyakant, Sudipto Das, and Amr El Abbadi. 2011. Big data and cloud computing: Current state and future opportunities. In *Proceedings of the 14th International Conference on Extending Database Technology*, EDBT/ICDT '11, 530–533, New York, NY, USA, ACM.
4. Andrikopoulos, V., S. Benbernou, and M.P. Papazoglou. 2012. On the evolution of services. *IEEE Transactions on Software Engineering*, 38(undefined): 609–628.
5. Andrikopoulos, V. and P. Plebani. 2011. Retrieving compatible web services. *2011 IEEE International Conference on Web Services (ICWS 2011)*, 00(undefined): 179–186.
6. Autili Marco, Paola Inverardi, Alfredo Navarra, and Massimo Tivoli. 2007. Synthesis: A tool for automatically assembling correct and distributed component-based systems. In *Proceedings of the 29th International Conference on Software Engineering*, ICSE '07, 784–787. Washington, DC, USA, IEEE Computer Society.
7. Bauer, Florian, and Martin Kaltenbock. 2011. Linked open data: The essentials. Edition mono/monochrom.
8. Bellini, P., M. Benigni, R. Billero, P. Nesi, and N. Rauch. 2014. Km4city ontology building vs data harvesting and cleaning for smart-city services. *Journal of Visual Languages and Computing* 25(6): 827–839.
9. Biem, Alain, Eric Bouillet, Hanhua Feng, Anand Ranganathan, Anton Riabov, Olivier Verscheure, Haris Koutsopoulos, and Carlos Moran. 2010. IBM infosphere streams for scalable, real-time, intelligent transportation services. In *Proceedings of the 2010 ACM SIGMOD International Conference on Management of Data*, SIGMOD '10, 1093–1104, New York, NY, USA, ACM.
10. Bischof, Stefan, Athanasios, Karapantelakis, and Cosmin-Septimiu, Nechifor, Amit P. Sheth, Alessandar Mileo, and Payam Barnaghi. 2014. Semantic modelling of smart city data. https://www.w3.org/2014/02/wot/papers/karapantelakis.pdf.
11. Bizer, Christian, Tom Heath, and Tim Berners-Lee. 2009. Linked data-the story so far. *Semantic Services, Interoperability and Web Applications: Emerging Concepts*.
12. Bowerman, B., J. Braverman, J. Taylor, H. Todosow, and U. Von Wimmersperg. 2000. The vision of a smart city. In *2nd International Life Extension Technology Workshop, Paris* 28.
13. Brogi, Antonio, Javier Cmara, Carlos Canal, Javier Cubo, and Ernesto Pimentel. 2007. Dynamic contextual adaptation. *Electronic Notes in Theoretical Computer Science* 175(2): 81–95.
14. Brogi, Antonio and Razvan Popescu. 2006. Automated generation of bpel adapters. In *Proceedings of the 4th International Conference on Service-Oriented Computing*, ICSOC'06, 27–39. Springer, Berlin, Heidelberg.
15. Cámara, Javier, José Antonio Martín, Gwen Salaün, Javier Cubo, Meriem Ouederni, Carlos Canal, and Ernesto Pimentel. 2009. Itaca: an integrated toolbox for the automatic composition and adaptation of web services. In *2009 31st. International Conference on Software Engineering. ICSE2009. May 16–24. Vancouver, Canada. Proceedings*, 627–630. IEEE Computer Society.
16. Canal, Carlos, Pascal Poizat, and Gwen Salaün. 2008. Model-based adaptation of behavioral mismatching components. *IEEE Transactions on Software Engineering* 34(4): 546–563.
17. Consoli, S., M. Mongiovic, A.G. Nuzzolese, S. Peroni, V. Presutti, R. Diego Reforgiato, and D. Spampinato. 2015. A smart city data model based on semantics best practice and principles. In *Proceedings of the 24th International Conference on World Wide Web*, WWW '15 Companion, 1395–1400. New York, NY, USA, ACM.
18. Cubo, Javier, and Ernesto Pimentel. 2011. *DAMASCo: A Framework for the Automatic Composition of Component-Based and Service-Oriented Architectures*, 388–404. Springer, Berlin, Heidelberg.

19. Cuzzocrea, Alfredo, Il-Yeol Song, and Karen C. Davis. 2011. Analytics over large-scale multidimensional data: The big data revolution! In *Proceedings of the ACM 14th International Workshop on Data Warehousing and OLAP*, DOLAP '11, 101–104. New York, NY, USA, ACM.
20. Demirkan, Haluk, and Dursun Delen. 2013. Leveraging the capabilities of service-oriented decision support systems: Putting analytics and big data in cloud. *Decision Support Systems* 55(1): 412–421.
21. European Commission. 2016. Making Big Data work for Europe. http://ec.europa.eu/digital-agenda/en/big-data.
22. Hashem, Ibrahim Abaker Targio, Ibrar Yaqoob, Nor Badrul Anuar, Salimah Mokhtar, Abdullah Gani, and Samee Ullah Khan. 2015. The rise of big data on cloud computing. Review and open research issues. *Information Systems* 47: 98–115.
23. Hollands, Robert, G. 2008. Will the real smart city please stand up? intelligent, progressive or entrepreneurial? *City* 12 (3): 303–320.
24. Lambda Architecture Net. 2016. Lambda Architecture. http://lambda-architecture.net/.
25. Manin, B. 1997. *The Principles of Representative Government*. Cambridge University Press.
26. Martin, J.A., F. Martinelli, and E. Pimentel. 2012. Synthesis of secure adaptors. *The Journal of Logic and Algebraic Programming*, 81(2):99–126. Formal Languages and Analysis of Contract-Oriented Software (FLACOS'10).
27. Marz, Nathan, and James Warren. 2015. *Big Data: Principles and Best Practices of Scalable Realtime Data Systems*, 1st ed. Greenwich, CT, USA: Manning Publications Co.
28. Motahari Nezhad, Hamid Reza, Boualem Benatallah, Axel Martens, Francisco Curbera, and Fabio Casati. 2007. Semi-automated adaptation of service interactions. In *Proceedings of the 16th International Conference on World Wide Web*, WWW '07, 993–1002. New York, NY, USA, ACM.
29. Motahari Nezhad, Hamid Reza, Guang Yuan Xu, and Boualem Benatallah. 2010. Protocol-aware matching of web service interfaces for adapter development. In *Proceedings of the 19th International Conference on World Wide Web*, WWW '10, 731–740. New York, NY, USA, ACM.
30. Papazoglou, Mike P. 2008. The challenges of service evolution. In *Proceedings of the 20th International Conference on Advanced Information Systems Engineering*, CAiSE '08, 1–15. Springer, Berlin, Heidelberg.
31. Schaffers, Hans, Annika Sällström, Marc Pallot, José M. Hernández-Muñoz, Roberto Santoro, and Brigitte Trousse. 2011. Integrating living labs with future internet experimental platforms for co-creating services within smart cities. In *Concurrent Enterprising (ICE), 2011 17th International Conference on*, 1–11. IEEE.
32. Talia, Domenico. 2013. Clouds for scalable big data analytics. *Computer* 46(5): 98–101.
33. Toshniwal, Ankit, Siddarth Taneja, Amit Shukla, Karthik Ramasamy, Jignesh M. Patel, Sanjeev Kulkarni, Jason Jackson, Krishna Gade, Maosong Fu, Jake Donham, Nikunj Bhagat, Sailesh Mittal, and Dmitriy Ryaboy. 2014. Storm@twitter. In *Proceedings of the 2014 ACM SIGMOD International Conference on Management of Data*, SIGMOD '14, 147–156. New York, NY, USA, ACM.
34. W3C Incubator Group Report. 2011. Semantic sensor network XG final report. http://www.w3.org/2005/Incubator/ssn/XGR-ssn.
35. Wetzstein, Branimir, Dimka Karastoyanova, Oliver Kopp, Frank Leymann, and Daniel Zwink. 2010. Cross-organizational process monitoring based on service choreographies. In *Proceedings of the 2010 ACM Symposium on Applied Computing*, SAC '10, 2485–2490. New York, NY, USA, ACM.
36. Zulkernine, F. P. Martin, Y. Zou, M. Bauer, F. Gwadry-Sridhar, and A. Aboulnaga. 2013. Towards cloud-based analytics-as-a-service (claaas) for big data analytics in the cloud. In *2013 IEEE International Congress on Big Data*, 62–69.

A Case for IoT Security Assurance

Claudio A. Ardagna, Ernesto Damiani, Julian Schütte and Philipp Stephanow

Abstract Today the proliferation of ubiquitous devices interacting with the external environment and connected by means of wired/wireless communication technologies points to the definition of a new vision of ICT called Internet of Things (IoT). In IoT, sensors and actuators, possibly embedded in more powerful devices, such as smartphones, interact with the surrounding environment. They collect information and supply it across networks to platforms where IoT applications are built. IoT services are then made available to final customers through these platforms. Needless to say, IoT scenario revolutionizes the concept of security, which becomes even more critical than before. Security protection must consider millions of devices that are under control of external entities, freshness and integrity of data that are produced by the latter devices, and heterogeneous environments and contexts that co-exist in the same IoT environment. These aspects make the need of a systematic way of assessing the quality and security of IoT systems evident, introducing the need of rethinking existing assurance methods to fit the IoT-based services. In this chapter, we discuss and analyze challenges in the design and development of assurance methods in IoT, focusing on traditional CIA properties, and provide a first process for the development of continuous assurance methods for IoT services. We also design a conceptual framework for IoT security assurance evaluation.

C.A. Ardagna (✉)
Dipartimento di Informatica, Università Degli Studi di Milano, Milano, Italy
e-mail: claudio.ardagna@unimi.it

E. Damiani
Etisalat British Telecom Innovation Center, Khalifa University of Science, Technology and Research, Abu Dhabi, UAE
e-mail: ernesto.damiani@kustar.ac.ae

J. Schütte · P. Stephanow
Fraunhofer Institute for Applied and Integrated Security, Garching Near Munich, Germany
e-mail: julian.schuette@aisec.fraunhofer.de

P. Stephanow
e-mail: philipp.stephanow@aisec.fraunhofer.de

B. Di Martino et al. (eds.), *Internet of Everything*, Internet of Things,
https://doi.org/10.1007/978-981-10-5861-5_8

1 Introduction

On 21st October 2016, a security incident of the Internet of Things (IoT) gained wide public attention: an attack tool named *Mirai* used IoT devices to launch a massive Distributed Denial-of-Service (DDoS) attack against Dyn,[1] which affected the availability of large online platforms such as Twitter, Amazon, Tumblr, Reddit, Spotify and Netflix.[2] More than 100,000 devices had been taken over by the botnet and used to attack the most prominent services of the Internet, reportedly at volumes of up to 1.2 Tbps. The vulnerabilities exploited to compromise IoT devices such as cameras and network video recorders were simple and well-known [1]: firstly, the infected devices were shipped with publicly known factory-default administration accounts. Secondly, the default configuration for OpenSSH daemons running on these devices allowed TCP forwarding. Although this configuration bug has been reported since 2004 [2], a large number of IoT devices deployed in 2016 still have this vulnerability, allowing attackers to misuse the devices as SOCKS proxies to carry out DDoS attacks. Despite unprecedented in scale, this incident reveals only a small fraction of security issues and challenges that IoT face.

IoT systems form the technical backbone to delivery IoT services. While IoT services are not based on revolutionary new technologies, such examples show that the way in which they are deployed imposes fundamentally new threats and consequently challenges on assuring their functionality and security.

First, the Internet of Things comprises a vast amount of connected devices with specific purposes, such as cameras, sensors, and actuators. For the majority of these devices, cost pressure, short time to market, and their limited functionality prohibit extensive development of security mechanisms. This intrinsic insecurity results in large-scale deployments of connected devices with homogeneous platforms and mostly no remote update capabilities—an Internet of Unpatchable Things serving as an ideal target for long-lasting botnets and easy entry points into otherwise protected networks.

Second, ownerships and responsibilities are scattered in IoT services. The proposed example stresses clearly how no entity in the chain, from device manufacturing over roll-out to operations, is in charge of preventing third parties from implementing large-scale attacks on the IoT infrastructure. In contrast to enterprise IT where clear perimeter protection is in place and expert staff is in charge of running operations securely and reliably, IoT devices are often found in private households and public places, and operate without any maintenance for years. Vulnerabilities and backdoors are rarely discovered at all, and in case they are, it is unlikely that owners have the knowledge, resources, and motivation to fix them.

Third, risk management is also fundamental in IoT services. Security-related risks are the most obvious and urgent ones, but the combination of large-scale deployed homogeneous platforms and scattered responsibilities also lead to legal risks, privacy risks, and risks of violating defined business processes. For instance, if business

[1] https://dyn.com/blog/dyn-analysis-summary-of-friday-october-21-attack/.

[2] https://www.theguardian.com/technology/2016/oct/26/ddos-attack-dyn-mirai-botnet.

processes are based on data that are collected by IoT devices, which are not under control of the enterprise, how reliable is information gained from the devices' data? Current approaches rely on the law of large numbers and assume that if sensor data are manipulated at all, only a small fraction of devices will be affected. This assumption however does not hold in a real IoT system as indicated by attacks and scenarios described above.

As of today, there is a lack in systematic assessing the quality and security of IoT systems. Consequently, users willing to embrace IoT systems in their business are facing the following questions:

- How can integrity of data provided by IoT systems and the reliability of business decisions taken on that basis be assessed and controlled?
- When given the choice between IoT systems, how can users determine which one matches best their requirements on security and quality?
- How can the risks of data loss, privacy breaches, and resulting liabilities be controlled—when hosting IoT systems as well as when making use of them?

Assurance methods provide the answers to these questions: these methods aim to validate whether a generic service adheres to a set of (security) requirements, thereby increasing service consumers' trust that the service is behaving as expected, and enabling comparability. Yet, since IoT systems rely on distributed and heterogeneous components and devices, often deployed over heterogeneous infrastructures, manually evaluating consumers' requirements satisfaction is *not* feasible. Furthermore, an IoT system's attributes may change over time in a way that is not predictable or detectable by a consumer. Examples are configuration changes, patches applied to service components or rather frequently expected failure of low-end embedded systems, i.e. IoT devices on the perception layer, such as sensors.

Developing assurance methods for IoT systems therefore requires an approach capable of *continuously*, that is, automatically and repeatedly detecting ongoing changes and assessing their impact on consumer security requirements. Furthermore, providing evidence that an IoT system satisfies a security requirement at a certain point in time paves the way for security audits and security certification of IoT systems, and requires a compositional approach where local evidence on the status of given objects is composed to provide process-wide claims. Recent research initially provided surveys on security challenges of IoT [3–7]. Other work proposed security mechanisms on how to address these challenges, such as [8–14]. However, they fall short of developing methods to continuously evaluate whether an IoT system satisfies a set of security requirements over time. This scenario points to the need of rethinking the design, development and deployment of existing assurance methods [15].

In this chapter, we propose a framework to support research activities aimed to design and implement methods enabling continuous and compositional security assurance of IoT systems. To this end, after introducing the concepts of IoT and security assurance (Sect. 2), we discuss general, novel security requirements of IoT systems and outline corresponding approaches of current research (Sect. 3). Then, we discuss challenges of continuously assuring security properties of IoT systems

(Sect. 4), proposing some guidelines on the development of security assurance methods for IoT systems and the design of a conceptual framework for IoT security assurance evaluation.

2 Background

Both the *Internet of Things* and *security assurance* terms lack of a precise and commonly accepted definition. In this section, we establish common ground by describing both terms in the context of this chapter.

2.1 Internet of Things

The Internet of Things (IoT) is a term used today to describe the increasing interweaving of machines, their Operational Technology (OT), Information Technology (IT), their physical environment, and the user. A somewhat more formal definition is proposed by the ISO/IEC where IoT is an "[..] *infrastructure of interconnected objects, people, systems and information resources together with intelligent services to allow them to process information of the physical and the virtual world and react*" [16].

The Internet of Things refers to the interconnection of technical objects for the sake of smart and data-driven applications interacting with the physical world. From a technical perspective, IoT is rather an evolution, driven by miniaturization of powerful embedded platforms, the development of lightweight protocols such as MQTT, CoAP, and LWM2M, and the rise of data-heavy cloud applications [17]. From a societal perspective, however, IoT is about to foster a revolution: the way users interact with applications is dramatically changed by mobile devices and embedded systems ubiquitously integrated into the physical world. The rapidly growing amount of data collected by these devices enables applications to apply advanced data processing to make predictions and take decisions with impact in the physical world.

Although its undebatable advantages and promises, a canonical understanding of IoT allowing for an unambiguous usage of the term is not available [18], resulting in a scenario where terms such as IoT system, IoT service, and IoT application are inconsistently used. In this chapter, we use the term IoT(-based) service describing a service delivered through an IoT system or environment. The origins of IoT lie in the concept of pervasive computing (the term ubiquitous computing is often used interchangeable), coined by Mark Weiser in this article "The computer for the 21st century": "The most profound technologies are those that disappear. They weave themselves into the fabric of everyday life until they are indistinguishable from it." Weiser [19] The omnipresent interaction with hidden computers integrated into everyday objects has been guiding several research areas since then, ranging from human-machine-interaction over networking to security. Commercialization was picking up with the advent of smartphones, cheap but powerful programmable embedded platforms such as the Raspberry PI or the Arduino, and affordable smart home applications. In this context the term Internet of Things became increasingly popular.

Today, despite the lack of an authoritative definition, we can sufficiently precisely describe instances of *IoT services* by their properties. First they increase the complexity of ICT systems extending them with a large number of highly distributed and heterogeneous software, hardware, network and sensing components. Second they consist of products and services that rely on distributed and heterogeneous components and devices, often deployed over heterogeneous infrastructures which operate under non-centralized ownership and control. Finally, they pose strong requirements on performance and battery consumption. Such properties pave the way for a multitude of novel applications, optimization, and use cases [17], such as: *highly resilient production environments*, where data collected from a company's IT and OT systems can be used to generate advanced risk profiles; *smart factories and real-time supply chains*, where data sources and feedback loops along the whole supply chain from manufacturers to consumers allow for highly flexible production processes and timely optimized manufacturing of individual demands; *environment protection and improved public safety*, where IoT services can revise and optimize processes in metropolitan areas, such as reducing traffic congestion and pollution; *improved life quality of individuals*, where services are directly provided to individuals, such as comprehensive health monitoring, diagnostics and even medication.

Al-Fuqaha et al. [20] made a step forward defining a 5-layered architecture for IoT systems, also including the concepts of IoT service and application. At the first layer (object layer) are physical sensors collecting and processing information. On top of it (object abstraction layer) data produced by sensors are securely transferred to middleware (service management layer) binding an *IoT service* with its requester. The middleware supports IoT application programmers to work with heterogeneous objects abstracting from the specific hardware platform. The services requested by final customers are then delivered at application layer providing the entry points to them for accessing smart services. Finally, IoT system activities and services are monitored at the top layer of the architecture (business layer). It is within this layer where assurance methods are deployed to evaluate the behavior of a system and support decision making.

2.2 Security Assurance

In line with standards, *security assurance* can be defined as a way to gain confidence of correct behavior of a system [21]. *Security assurance* is justifiable confidence that a system will consistently demonstrate one or more security properties and thus satisfies its security requirements, regardless of failures and attacks [15, 21]. This *confidence* is based on evaluating evidence, that is, observable information of the system.

Security assurance methods describe how to collect and evaluate evidence to determine satisfaction of security properties. While some assurance methods can be used throughout the whole life cycle of a system, such as external reviews, others can be mapped to specific phases of a system's life cycle [22]:

- Assuring requirement collection: are the collected security requirements complete, sound, and consistent?
- Assuring system design: does the system design satisfy defined security requirements?
- Assuring secure implementation: does the implementation of the system meets its security requirements?

Providing assurance of implementation covers assuring satisfaction of security requirements at development as well as at deployment time. During development, *static assurance methods* which do not require executing the system can be used. Examples for such methods are security testing techniques such as static code analysis or code review. At deployment time, dynamic assurance methods are needed to check fulfillment of security requirements while a system is running. Among them, *test-based assurance methods* [15] produce evidence by controlling some input to the system and evaluating the output, such as for instance calling an IoT service's CoAP API and checking responses; *monitoring-based assurance methods* [15] use monitoring data as evidence collected from components involved in the delivery of, for instance, an IoT service. Monitoring-based methods, while in general more costly, are often used in scenarios where a testing-based approach only provides insufficient evidence or is usually forbidden, for instance, when exploiting a vulnerability as part of a security testing technique. Finally, *hybrid assurance methods* combine test-based and monitoring-based evidence since monitoring or testing alone can only cover parts of an IoT service's behavior.

While assuring security of IoT services is evidently indispensable, design and implementation of corresponding security assurance methods raise several research challenges. In an attempt to systematically approach security assurance for IoT, we identified the following concepts which need to be considered when developing security assurance methods for IoT services.

Untrusted endpoint operation. IoT services build on a large number of heterogeneous embedded devices acting as data sources. Many are resource constrained, untrusted and unreliable and are operated by various owners, usually non-professionals. As a consequence, even basic security requirements such as regular software updates of endpoints do not hold.

Transparent service composition and delivery. From a service customer's point of view, an IoT service hides its distributed composition, that is, which and how components involved in delivery of an IoT service collaborate is not discernible for a service customer. Since multiple, potentially complex systems, such as cloud services, take part in providing an IoT service, establishing a level of trust in the overall IoT service, as well as specific outputs it delivers, depicts a hard challenge.

This points to a scenario where local claims on specific objects and devices are composed to provide a process-wide assurance evaluation of composite services.

Data-centric applications. While devices involved in IoT service delivery serve as sensors or actuators, applications make use of advanced algorithms to analyze data collected by the devices, for instance to infer environmental conditions, predict traffic situations, control smart homes or adapt manufacturing processes. This has various implications on security and privacy: while in the past the major focus of security measures was endpoint and perimeter security, this focus now shifts to trustworthiness of data processing chains. Also, as data collection and its usage are decoupled, it becomes hard to assess privacy properties of data collected by a specific device when used in varying contexts, that is, processed by different applications and combined with other data sources.

Data Quality. In a complex IoT service, data quality can be considered from two opposite points of view. On one side, data quality is at the basis of an accurate and precise security assurance method; on the other side, assurance methods must be used to evaluate the quality of data produced by IoT services.

Limited Resource and Heterogeneous Devices. IoT environments are composed of billion of devices and sensors, several different networks, and data centers. Each of these components has its own peculiarities, requirements, and limitations. A proper assurance method for IoT must accomplish all of these components being flexible, adaptive, and dynamically configurable.

Decentralization and Geographical Distribution. IoT environments are decentralized and geographically distributed by nature. Assurance methods must adapt to these peculiarities going even beyond the borders introduced by traditional distributed systems.

3 Challenges to IoT Security Assurance

In this section, we present challenges of continuously assuring security properties of IoT services. To that end, we draw on the classical security goals—Confidentiality, Integrity, and Availability—to describe security requirements of IoT service, also pointing out mechanisms identified by recent research [23, 24] to meet these requirements. Drawing on these requirements and proposals, we discuss challenges to continuous security assurance of IoT services.

3.1 Confidentiality

Confidentiality refers to the property of an IoT service that any information in the context of the service is only disclosed to authorized parties. This includes information provided as input to the service, information processed and stored by the service as well as information produced as output by the service. Since IoT services are composed of different independent systems interacting with each other, processing information also includes communications between systems involved in service delivery.

3.1.1 Providing Confidentiality

To provide confidential communications between resource constrained devices, the constrained application protocol (CoAP) advocates using DTLS [25]. However, Raza et al. [9] argue that the DTLS protocol was originally designed for communications where the message length was not critical rendering DTLS unsuitable for resource-constrained devices [9]. They therefore propose an adaption providing a more resource efficient variant called *CoAPs Lithe*. In a different line of work, Raza et al. [12] present an adaptation of the IPv6 Over Low-power Wireless Personal Area Networks (6LoWPAN) [26], where they extend 6LoWPAN to use IPsec [27].

Some work aims to provide confidentiality of information at rest, in particular information processed and stored by resource constrained devices at the perception layer. Bagci et al. [8] propose an approach to efficiently store confidential data on sensor nodes, while still allowing data processing on nodes. They argue that simply encrypting data before usage as proposed by, for instance, Bhatnagar and Miller [28] and Ren et al. [29] hinder in-network data processing. Further, Dofe et al. [10] use basic message permutations to make hardware attacks, such as side-channel attacks, on resource constrained devices harder.

3.1.2 Assuring Confidentiality

Assuring the confidentiality of data at rest, that is, temporarily or persistently stored on a IoT service's component, and data in transit, that is, data transferred between different Iot service components, is hard since we are facing a heterogeneous set of components involved in service delivery, which may change over time. Thus we can assume that there are various security mechanisms implemented by different applications in place aiming to provide data confidentiality, such as for instance encryption of block devices using *dm-crypt* provided by Linux kernel and encryption of data in transit using CoAP over DTLS [25] or Lithe [9].

Applying the challenges of accuracy, precision and completeness to confidentiality translates to the following: each mechanism an IoT service deploys to provide confidentiality needs a suitable assurance method, leading to a set of required assurance methods. This challenge further aggravates when considering that an IoT service's composition may change over time, also altering the set of deployed mechanisms to provide confidentiality. Thus we need to ensure that for a particular service composition at time t the suitable set of assurance methods is selected and deployed. Yet, if we cannot be perfectly sure that we have a set of assurance methods at hand to allow for correct, precise, and complete reasoning about IoT service's confidentiality, then we need to devise a measure to describe our confidence as to what degree the result of the assurance method is correct, precise, and complete.

Naturally, leaking the results of assessing whether an IoT service is satisfying the security property confidentiality can cause serious security issues. Thus a suitable security model for assurance methods is required. This challenges is further exacerbated by a particular data processing strategy which aims to process sensitive and privacy-related data locally, that is, close to the sensor and not be forwarded to some external application which provides data analytics within the IoT service [3]. For this reason, a conflict exists between the intrinsic need of assurance methods to provide their results beyond the local data processing layer and the strategy to keep sensitive data local. To solve this conflict, mechanisms are needed to decide whether results produced by assurance methods pose a security risk to the IoT service and, if so, how to pre-process the data locally to lower the risk below an acceptable limit while still providing sufficient information to allow for reasoning about confidentiality of the IoT service.

3.2 *Integrity*

Integrity refers to the property of an IoT service to only permit authorized parties modifications of any information involved in IoT service delivery. Data and communication integrity are among the most critical security properties of any distributed and large scale systems [30]. The need of strong integrity guarantees is at the core of trustworthy IoT services where heterogeneous systems and devices are used providing high volumes of data, which serve as input to advanced data analysis. Maliciously altered data can lead to incorrect results of data analysis, paving the way for novel attack scenarios such as adversarial machine learning [31, 32].

3.2.1 Providing Integrity

Implementing mechanisms to provide data and communications integrity of an IoT service is a complex problem which requires configurations and algorithms working at different layers of the IoT service stack. Existing techniques for data integrity (and confidentiality) are often based on encryption techniques, such as Transport Layer Security (TLS) or Internet Protocol Security [30]. A major problem with these techniques in the IoT scenario is in the distribution of keys and certificates. Traditional solutions like the one based on public key infrastructures or proprietary hardware/firmware with hard-coded credentials are difficult to apply in IoT environments. The first approach does not fit large deployments, while low flexibility of approaches based on hard-coded credentials make it difficult to manage scenarios in which credentials are compromised and need to be updated. In [30], a specific approach based on Generic Bootstrapping Architecture (GBA) technology and

Authentication and Key Agreement (AKA) protocol can be used to support communication and data security. Liu et al. [33] provide an analysis on authenticator-based techniques for data integrity verification in IoT. The analysis starts from the claim that data integrity, a fundamental aspect for data security, is inherently different in IoT scenarios that merge the peculiarities of cloud and big data environments. Data are in fact dynamic in nature and are composed of huge amount of very small chunks that are frequently updated. This scenario points to the need of techniques for the verification of dynamic data. Data integrity solutions must then support three main aspects efficiency, security, and scalability/elasticity. Newe et al. [34] deal with the problem of verifying data integrity in IoT using hardware implementations of cryptographic hash algorithms (e.g., ASICs and FPGAs hardware platforms), and propose an efficient high-speed FPGA implementation of the newly selected hash algorithm, SHA-3. This approach aims to satisfy the need of efficiency and near real-time data integrity checking.

Other techniques (e.g., [35]) propose the adoption of block chain-based approaches that have been used for integrity of crypto-currencies. Moreover, software-based attestation protocols [14] have also been used to verify the integrity of a given smart meter and its data. In sensor networks, different approaches to data integrity have been introduced [36]. Among them, CoAPs Lithe [12] and 6LoWPAN/IPsec [12] have been discussed in the previous section.

3.2.2 Assuring Integrity

Assurance methods for data and communication integrity should prove trustworthiness of data along the whole IoT service delivery, clearly identifying liability for erroneous or malformed data and communication distribution. Further, assurance methods should keep the measurement overhead of integrity verification manageable in practice, especially with regard to devices with limited capabilities. Being based on cryptographic approaches, assurance methods should support checking compliance of traditional integrity solutions for unconstrained devices, as well as lightweight counterparts for sensors and resource-constrained devices.

Moreover, continuous assurance methods should provide the ability to check integrity in near real time, promptly detecting non-compliance of an IoT service. They need also support the verification of data integrity at all layers of the IoT service, allowing the integrate heterogeneous evidence with different levels of accuracy and precision into their evaluation process.

A posteriori verification of assurance methods' results is also critical. Yet this goal is difficult to achieve in an IoT service where the rate of node failures and events (e.g., node join/leave) is expectedly rather frequent. A collaborative way of verifying data integrity is thus needed to compensate for the high rate of changes in delivery of an IoT service.

3.3 Availability

Availability refers to the probability that an IoT service is operating as expected at any given moment, ready to deliver its service to the service customer. In the context of industrial applications of IoT services. Sadeghi et al. [3] point out that availability is one of the most important security goals. The reason for this stems from potential loss of productivity and thus revenue in case an industrial production system process is delayed or even postponed as a result of IoT services' unavailability.

Besides industrial applications of IoT services, there are further exemplary scenarios which help illustrating that IoT services violating necessary availability requirements can entail undesired consequences. Consider, for example, a medical IoT service which serves to measure blood glucose level of diabetes patients at defined intervals and administer injections of insulin if required. The availability of this IoT service is vital since missing insulin doses can seriously harm the patient.

3.3.1 Providing Availability

When considering the composition of an IoT service, frequent failure of low-end embedded systems at the perception layer is rather expected. However, all devices of an IoT service failing simultaneously is rather unlikely. Therefore, we can expect *partial* failures, a standard notion used in distributed systems, leading to the requirement of fault tolerance [37]: Whenever components or communications between components involved in delivery of an IoT service fail, the IoT service should be capable of tolerating this failure and continue delivering the service as expected by the service customer.

There are mature solutions to masking failures within distributed systems by redundancy, that is, information redundancy, time redundancy, as well as physical redundancy [37]. However, with regard to IoT services we face interaction between different heterogeneous, potentially distributed systems, often deployed over heterogeneous infrastructures. Developing mechanisms to ensure high availability of IoT services thus becomes a difficult challenge.

3.3.2 Assuring Availability

A naive approach to assure availability of an IoT service consists of decomposing the service and choose suitable assurance methods to check availability of each component. This implies that composition of an IoT services is known at the time when an assurance method is applied. Alternatively, we may neglect actual service compositions by defining an IoT service's availability strictly based on the service customer's view. In this case, the service is available if it behaves as expected when the customer is interacting with it. Yet both perspectives appear rather simplistic when considering that an IoT service's components and behavior may change over time and, in particular, when factoring in that failures of low-end devices and communications at the perception layer are somewhat expected. Therefore, when determining the availability of an IoT service, we have to take uncertainty about assurance methods' results into account. Therefore, a methodology is required to describe to what extent we consider our statements about the availability of an IoT service to be correct as well as complete.

Another challenge to be mastered when assuring availability lies in overhead incurred by assurance methods' application. Naturally, continuously validating availability of an IoT service comes at a price, that is, will incur overhead on the IoT service's components. In context of resource-constrained devices of IoT services, such as for instance sensors for environmental monitoring, continuously assuring availability becomes a hard challenge. Thus, in order not to compromise availability of an IoT service by assuring its availability, methods are required to efficiently deploy required availability checks while retaining necessary accuracy of results.

4 Developing Continuous Security Assurance Methods

This section describes a process supporting the development of assurance methods that continuously check whether an IoT service complies with a defined set of security requirements, and provides a first framework for IoT security assurance evaluation.

4.1 IoT Security Assurance Process

While it is conceivable to continuously assure security properties at any stage of an IoT service's life cycle, our focus lies on designing security assurance methods to foster secure implementation, that is, checking compliance of an IoT service with security requirements both at development and deployment time. The proposed process is composed of five main stages as discussed in the following and shown in Fig. 1.

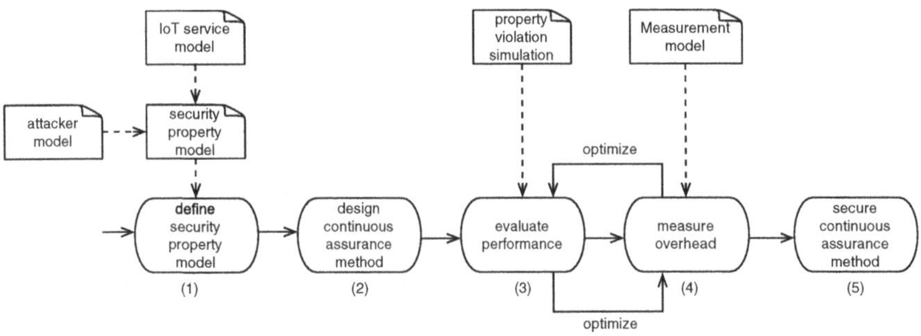

Fig. 1 Development stages of continuous assurance methods for IoT services

(1) **Define security property model**. Security requirements derived from, for instance, NIST SP 800-53 [38] or ISO27001:2013 [39] are generic and often times inherently ambiguous, making automatic validation infeasible. Thus the support of a continuous verification process checking whether an IoT service satisfies a set of security requirements requires security property models that can be automatically evaluated, thereby bridging the *semantic gap*.

Defining an IoT service's security properties leads to the classical dilemma that assessing non-functional properties ideally implies that any possible state of the IoT service has to be checked in order to be certain that the property holds. Naturally, this is hardly applicable in practice. Modeling security properties thus requires a risk-based assessment, that is, needs to consider an IoT service's assets as well as the skills and resources of specific adversaries, that is, an attacker model.

(2) **Design assurance method**. The design of a proper assurance method and the selection of the proper technology (e.g., testing, monitoring) at its basis often depend on the specific IoT service and the considered security properties. Without a proper design, the effectiveness of assurance methods would substantially decrease, also hindering the soundness of the overall approach and the quality of the collected results.

When IoT services are considered, testing and monitoring techniques need then to be carefully managed to accomplish IoT requirements. IoT testing should depart from the traditional view of testing-based verification. IoT systems are a mix of technologies, components, and infrastructure with different life cycles, which do not fit the traditional way of testing systems a priori in a lab environment. By contrast, testing techniques can be used to verify specific components of IoT, enabling the evaluation of specific aspects of IoT systems. Concerning monitoring techniques, IoT systems pose strict requirements on their deployment. The assumption of a complete monitoring of the whole system cannot be assumed in IoT environments. Therefore, strategic deployments should be foreseen with optimized placement of monitoring probes.

(3) **Evaluate performance**. The accuracy of results produced by a specific assurance method depends on various factors, such as implementation, environment, usage of external tools, and the like. Without experimental evaluation, it is hard to make a statement on how well a specific continuous assurance method detects compliance requirements violations.

A *simulation* manipulates an IoT service to mock violations of security requirements an assurance methods aims to detect. Property violation simulations are essential to experimentally analyze the performance of a specific assurance methods, that is, how well does the method work in detecting security property violations? It establishes the ground truth to which results produced by assurance methods are compared. Simulation happens prior to the productive deployment of an assurance method, for example, during integration testing or staging of the IoT service.

(4) **Measure overhead**. The strict interpretation of *continuous* security assurance of IoT services is hardly applicable in practice, since uninterruptedly assessing security requirements' satisfaction can incur intolerable overhead. This challenge further aggravates when considering low-end, resource-constrained devices. There is then the need to implement a measurement methodology that permits to reason about the overhead incurred by repeatedly checking security properties of IoT services, especially when facing multiple, concurrent, continuous assessments.

Analogous to performance evaluation, assessing the overhead of a specific, continuous assurance method can be conducted prior to its deployment. When also taking performance evaluation into account, this allows to compare alternative assurance methods, as well as alternative method configurations, based on performance and overhead. This way the most suitable assurance method including its optimal configuration can be selected.

(5) **Secure assurance methods**. Mechanisms seeking to increase trust and transparency can leak critical information, which can be used by attackers to trace vulnerabilities of an IoT service. It is clear that results produced by assurance methods which aim to detect violations of security requirements can contain critical information. Thus, it is vital that the system implementing continuous security assurance of IoT services is trustworthy as well.

4.2 A Framework to Support IoT Security Assurance

Figure 2 shows a preliminary design of a conceptual framework to support IoT Security assurance evaluation. The core of the framework is the *Assurance Manager* that is responsible for assurance evaluation and management, including assurance of composite IoT processes. To this aim, the Assurance Manager builds on local assurance models, which are implemented as a set of assurance mechanisms based on testing and monitoring agents, and used to collect assurance evidence (local claims) on the status of a given object or sub-system under evaluation. On top of local assurance models, the Assurance Manager provides functionalities for compositional assurance, where process-wide claims are set up to drive the collection of local claims and their integration, according to a machine-readable composition model. The Assurance Manager can finally interface with IoT middlewares (e.g., SOFIA—http://sofia2.com/home_en.html) to support users and service providers in an assurance-aware deployment of their applications (e.g., helping them in defining the IoT hooks that will be used by the assurance mechanisms to collect evidence through the IoT middleware).

The *Claims Parser* is the component responsible for (semi-) automatically translating a set of assurance claims (either local claims or process-wide claims) and a model of the (composite) service under evaluation into a given set of specific configurations, which define the assurance mechanism behavior during assurance evaluation including how to connect them to the IoT hooks. Assurance claims specify the assurance controls (tester, monitors) that are needed to achieve them, as well as

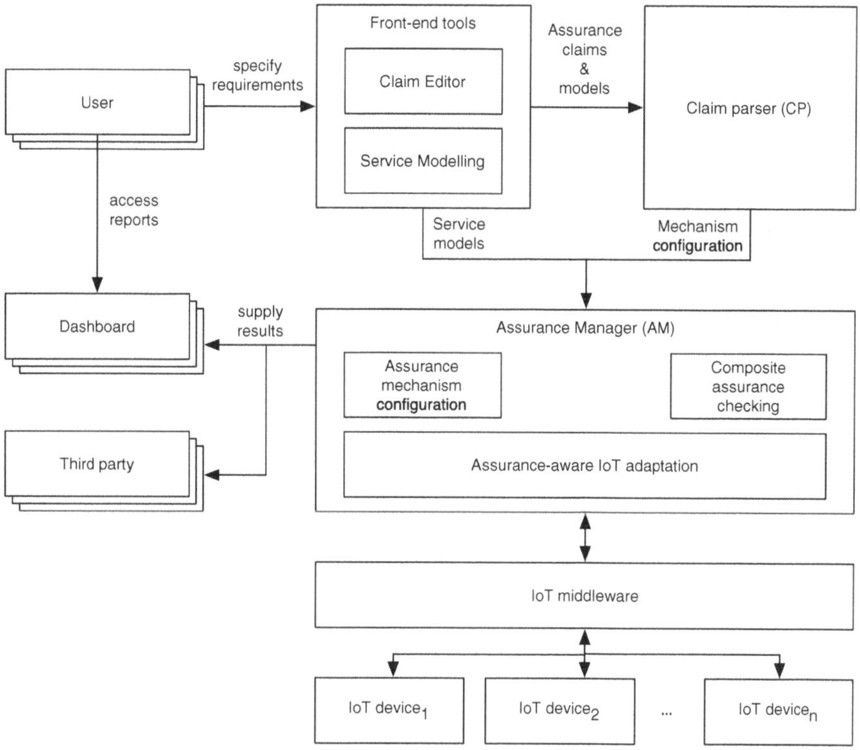

Fig. 2 Conceptual framework to support IoT security assurance evaluation

their configurations. Furthermore, assurance claims are used to select and configure assurance mechanisms with the Assurance Manager.

Upon selecting assurance mechanisms, Assurance Manager attaches the assurance mechanisms to the hooks for evidence collection and local/process-wide claim verification, using an IoT middleware. The framework must provide functionalities for managing assurance both at development and deployment time, managing the whole service lifecycle through configuring the Assurance Manager accordingly.

Finally, the Assurance Manager presents assurance results to users and third parties through the definition of a set of standardized interfaces with common syntax and semantics. Access to results can also be granted through dashboards to increase the usability of the tools and mechanisms. An IoT security assurance framework fosters the definition of a trustfully IoT environment able to attract critical services with strong security and privacy requirements, exposing public APIs for assurance management and for accessing the results of assurance evaluation.

5 Conclusions and Future Work

In this chapter, we discussed the challenges involved in assuring security properties of IoT services. To this end, we built on the three classic security goals—Confidentiality, Integrity, and Availability—to derive security requirements of IoT services. We then discussed research approaches that aim to meet these requirements and highlighted the new issues emerging when continuously assessing the behavior of security mechanisms. Finally, we laid out a 5-stage process guiding the development of assurance techniques to continuously check compliance of an IoT service against security requirements, and proposed a first design of an IoT security assurance framework.

As part of future work, we will investigate domain-specific scenarios, such as industrial IoT services and health care IoT services, to extract real-world security and privacy needs. We will then develop suitable continuous assurance techniques following the guidelines presented in this chapter.

Acknowledgements This work was partly supported by the program "piano sostegno alla ricerca 2015-17" funded by Università degli Studi di Milano.

References

1. Ezra Caltum and Ory Segal. SSHowDowN: Exploitation of IoT devices for Launching Mass-Scale Attack Campaigns. https://www.akamai.com/us/en/multimedia/documents/state-of-the-internet/sshowdown-exploitation-of-iot-devices-for-launching-mass-scale-attack-campaigns.pdf. Accessed 11 Oct 2016.
2. US-CERT/NIST. CVE-2004-1653. 2004. https://web.nvd.nist.gov/view/vuln/detail?vulnId=CVE-2004-1653. Aug, 2004. Accessed 11 2016.
3. Sadeghi, Ahmad-Reza, Christian Wachsmann, and Michael Waidner. 2015. Security and privacy challenges in industrial internet of things. In *Proceedings of the 52nd Annual Design Automation Conference (DAC)*, 54. ACM.
4. Abomhara, Mohamed and Geir M Køien. 2014. Security and privacy in the Internet of Things: Current status and open issues. In *International Conference on Privacy and Security in Mobile Systems (PRISMS)*, 1–8. IEEE.
5. Zhang, Zhi-Kai, Michael Cheng Yi Cho, Chia-Wei Wang, Chia-Wei Hsu, Chong-Kuan Chen, and Shiuhpyng Shieh. 2014. IoT security: ongoing challenges and research opportunities. In *2014 IEEE 7th International Conference on Service-Oriented Computing and Applications*, 230–234. IEEE.
6. Sato, Hiroyuki, Atsushi Kanai, Shigeaki Tanimoto, and Toru Kobayashi. 2016. Establishing trust in the emerging era of IoT. In *2016 IEEE Symposium on Service-Oriented System Engineering (SOSE)*, 398–406. IEEE.
7. Zhao, Kai, and Lina Ge. 2013. A survey on the internet of things security. In *Computational Intelligence and Security (CIS), 2013 9th International Conference on*, 663–667. IEEE.
8. Bagci, Ibrahim Ethem, Mohammad Reza Pourmirza, Shahid Raza, Utz Roedig, and Thiemo Voigt. 2012. Codo: Confidential data storage for wireless sensor networks. In *9th International Conference on Mobile Ad-Hoc and Sensor Systems (MASS)*, 1–6. IEEE.
9. Raza, Shahid, Hossein Shafagh, Kasun Hewage, René Hummen, and Thiemo Voigt. 2013. Lithe: Lightweight secure CoAP for the internet of things. *IEEE Sensors Journal* 13(10): 3711–3720.

10. Dofe, Jaya, Jonathan Frey, and Qiaoyan Yu. 2016. Hardware security assurance in emerging IoT applications. In *International Symposium on Circuits and Systems (ISCAS)*, 2050–2053. IEEE.
11. Raza, Shahid, Linus Wallgren, and Thiemo Voigt. 2013. SVELTE: Real-time intrusion detection in the Internet of Things. *Ad hoc networks* 11(8): 2661–2674.
12. Raza, Shahid, Simon Duquennoy, Joel Höglund, Utz Roedig, and Thiemo Voigt. 2014. Secure communication for the Internet of Things—a comparison of link-layer security and IPsec for 6LoWPAN. *Security and Communication Networks* 7(12): 2654–2668.
13. Lee, Jun-Ya, Wei-Cheng Lin, and Yu-Hung Huang. 2014. A lightweight authentication protocol for internet of things. In *2014 International Symposium on Next-Generation Electronics (ISNE)*, 1–2. IEEE.
14. Park, Haemin, Dongwon Seo, Heejo Lee, and Adrian Perrig. 2012. SMATT: Smart meter attestation using multiple target selection and copy-proof memory. In *Computer Science and its Applications*, 875–887. Springer.
15. Ardagna, Claudio Agostino, Rasool Asal, Ernesto Damiani, and Quang Hieu Vu. 2015. From security to assurance in the cloud: A survey. *ACM Computing Surveys (CSUR)*, 48(1): 2:1–2:50.
16. ISO/IEC JTC 1. 2014. Information Technology. Internet of things (iot). preliminary report.
17. B. Leukert et al. *IoT 2020: Smart and secure IoT platform*. IEC 2016. https://www.openstack.org/.
18. Minerva, Roberto, Abyi Biru, and Domenico Rotondi. 2015. *Towards a Definition of the Internet of Things (IoT)*. Torino, Italy: IEEE Internet Initiative.
19. Weiser, Mark. 1991. The computer for the twenty-first century. *Scientific American*, 6675.
20. Ala Al Fuqaha, Mohsen Guizani, Mehdi Mohammadi, Mohammed Aledhari, and Moussa Ayyash. 2015. Internet of things: A survey on enabling technologies, protocols, and applications. *IEEE Communications Surveys and Tutorials* 17(4): 2347–2376.
21. IATAC and DACS. 2007. Software security assurance: State of the art report (SOAR). http://www.dtic.mil/cgi-bin/GetTRDoc?Location=U2&doc=GetTRDoc.pdf&AD=ADA472363.
22. Beznosov, Konstantin, and Philippe Kruchten. 2004. Towards agile security assurance. In *Proceedings of the 2004 workshop on New security paradigms*, 47–54, ACM.
23. Misra, Sridipta, Muthucumaru Maheswaran, and Salman Hashmi. 2017. *Security challenges and approaches in internet of things*. Springer International Publishing.
24. Mahalle, Parikshit Narendra, and Poonam N. Railkar. 2015. *Identity management for internet of things*. River Publishers Series in Communications.
25. Shelby, Zach, Klaus Hartke, and Carsten Bormann. 2014. The constrained application protocol (CoAP). Technical report.
26. Montenegro, Gabriel, Nandakishore Kushalnagar, Jonathan Hui, and David Culler. 2007. Transmission of IPv6 packets over IEEE 802.15. 4 networks. Technical report.
27. Stephen Kent and Seo, Karen. 2005. Security architecture for the internet protocol. Technical report.
28. Bhatnagar, Neerja, and Ethan L. Miller. 2007. Designing a secure reliable file system for sensor networks. In *Proceedings of the 2007 ACM workshop on Storage security and survivability*, 19–24. ACM.
29. Wei Ren, Yi Ren, and Hui Zhang. 2008. Hybrids: A scheme for secure distributed data storage in wsns. In *IEEE/IFIP International Conference on Embedded and Ubiquitous Computing, 2008. EUC'08*, vol. 2, 318–323. IEEE.
30. Ericsson. 2016. Bootstrapping security-the key to internet of things access authentication and data integrity. Ericsson White paper, 284 23-3284. http://www.ericsson.com/res/docs/whitepapers/wp-iot-security.pdf.
31. Doug, J. 2011. Tygar. Adversarial machine learning. *IEEE Internet Computing* 15(5): 4.
32. Huang, Ling, Anthony D. Joseph, Blaine Nelson, Benjamin IP Rubinstein, and J.D. Tygar. 2011. Adversarial machine learning. In *Proceedings of the 4th ACM workshop on security and artificial intelligence*, 43–58. ACM.
33. Liu, Chang, Chi Yang, Xuyun Zhang, and Jinjun Chen. 2015. External integrity verification for outsourced big data in cloud and iot. *Future generation computer systems*, 49(C): 58–67.

34. Newe, Thomas, Muzaffar Rao, Daniel Toal, Gerard Dooly, Edin Omerdic, and Avijit Mathur. 2017. Efficient and high speed fpga bump in the wire implementation for data integrity and confidentiality services in the iot. In Postolache, Octavian Adrian, Subhas Chandra Mukhopadhyay, Krishanthi P. Jayasundera, and Akshya K. Swain (eds.). *Sensors for everyday life: Healthcare settings*, 259–285. Springer International Publishing.
35. Gaurav, Kumar, Pravin Goyal, Vartika Agrawal, and Shwetha Lakshman Rao. 2015. Iot transaction security. In *Proceedings of the 5th International Conference on the Internet of Things (IoT 2015)*.
36. Yick, Jennifer, Biswanath Mukherjee, and Dipak Ghosal. 2008. Wireless sensor network survey. *Computer Networks* 52(12): 2292–2330.
37. Tanenbaum, Andrew S., and Maarten Van Steen. 2007. *Distributed systems*. Prentice-Hall.
38. National Institute of Standards and Technology (NIST). 2013. Security and privacy controls for federal information systems and organizations. Special Publication 800: 53.
39. International Organization for Standardization (ISO). 2016. ISO/IEC 27001:2013 Information technology–Security techniques–Information security management systems–Requirements. https://www.iso.org/obp/ui/#iso:std:iso-iec:27001:ed-2:v1:en. Accessed 10 2016.

Study on IP Protection Techniques for Integrated Circuit in IOT Environment

Wei Liang, Jing Long, Dafang Zhang, Xiong Li and Yin Huang

Abstract The growth of electronic chip technique has led to frequent occurrence of intellectual property (IP) disputes. It seriously affects rapid and healthy development of semiconductor industry. To address the disputes, many IP protection methods are proposed in these years, such as IP watermarking. It is a novel technique to hide secrets in IP core to prove original ownership. This chapter focuses on two issues: how to hide secrets in IP circuit and how to authenticate IP ownership. Four types of IP watermarking methods will be concretely introduced in this chapter. (1) FPGA based IP watermarking technique. (2) FSM based IP watermarking technique. (3) DFT based IP watermarking technique. (4) Self-recoverable dual IP watermarking technique. The experiments show that the proposed schemes have low resource overhead by comparing with other schemes. Meanwhile the resistance to attacks of the watermark is encouraging as well.

1 Introduction

With the rapid development of internet of things (IOT), more and more transistors can be integrated into a single chip. The number has exceeded 10 billion in 2015. In Fig. 1, the complexity growth rate improves by 58% every year, but the productivity only grows by 21%. There is an increasingly deeper gap between chip-making capacity and design capacity. So, component based IP design method becomes prevalent due to its high efficiency [1]. IP reuse technology belongs to this type of design method.

W. Liang (✉)
Department of Software Engineering, Xiamen University of Technology,
Xiamen 361024, China
e-mail: idlink@163.com

J. Long · D. Zhang
College of Computer Science and Electronic Engineering, Hunan University, Changsha,
Hunan 410082, China

X. Li · Y. Huang
School of Computer Science and Engineering, Hunan University of Science and Technology,
Hunan 411201, Xiangtan, China

B. Di Martino et al. (eds.), *Internet of Everything*, Internet of Things,
https://doi.org/10.1007/978-981-10-5861-5_9

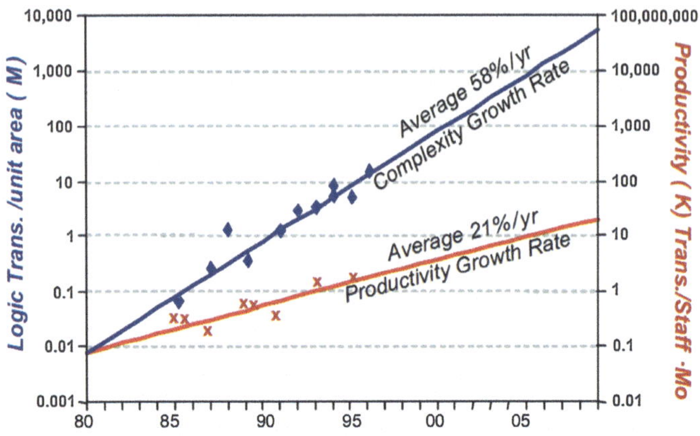

Fig. 1 Production gap between manufacture capacity and design capacity

It can save design cost, shorten design cycle, reduce market risk. Nowadays, it is a prevalent method in chip design.

Hardware device is the fundamental equipment in internet of things. So, the security of hardware integrated circuit should be guaranteed as well as software in internet of things. Nowadays, it is easy to reuse IP cores and manufacture various electronic products. The reused IPs may be misappropriated and utilized unauthorizedly for illegal profits. It directly leads to frequent occurrence of IP disputes every year [2]. Statistical data shows that financial loss caused by IP disputes achieves $50 billion every year [3]. Besides, it also brings damage to enterprise reputation and cooperative relationship. So, it is urgent to protect reused IPs from infringement. This subject has attracted many concerns in academia and semiconductor industry.

IP protection techniques can be classified into four categories: tagging, fingerprinting, watermarking and hardware encryption [4].

(1) Tagging. This technique places an electronical "label" into a chip for reliable and traceable identification. Marsh et al. [5] presented a tagging technique to protect Application Specific Integrated Circuit (ASIC) IP cores. A secure "label" identifying copyright information is placed into a chip. An "external receiver" is required to detect this label. But this method can only deter adversaries due to independence of the label. Besides, it might be damaged or removed. Another technique is physically unclonable function (PUF). It utilizes unique physical characteristics in IC manufacture to generate a radio frequency identification (RFID) "label", which is integrated into a chip to avoid cloning. The security is greatly enhanced, but expensive design cost and working environment of RFID have hindered its development [6].

(2) Fingerprinting. It makes different users get IPs with different identities. The uniqueness of IP fingerprinting realizes clear division of responsibility in IP disputes. But it will generate many IPs with the same functionality and technical index, but with different implementation. Lach et al. proposed an IP fingerprinting technique [7]. It divides a design into a set of parts that have the same characteristics. Each

part has several different implementations. IP module for embedding fingerprint is generated by combining different implementations of these parts. But this technique can only be realized at specific design level of very large scale integration (VLSI). Its application is limited due to the low resistance against collusion attacks.

(3) Watermarking. As a widely-used technique, watermarking is firstly applied in multimedia for copyright protection. In the field of VLSI, watermark is permanently stored in design as an invisible code for IP protection. Guneysu et al. [8] presented standard, protocol and design idea of reconfigurable digital watermark. Li et al. [9] concretely introduced development of IP watermarking and classified it into four categories at physical level, structural level, behavioral level and systematic level.

(4) Hardware encryption. Roy et al. [10] proposed an ending piracy of integrated circuits. A secret key is hidden into circuit. A chip cannot pass the test procedure and enter market if not activated. Besides, the authors also presented a bus based locking and unlocking scheme to protect hardware IPs. Although this technique increases hardware overhead (pins, area, etc.), it has good hiddenness and high security. But the protection is effective only in chip manufacture and test. The traceability after chip product being sold is not involved.

IP watermarking technique is a burgeoning interdisciplinary subject. It involves theories in various field, including microelectronics, signal processing, coding theory, cryptography, etc. So, it is of great significance and economical value to develop IP watermarking techniques. We have proposed four watermarking schemes to protect IP designs.

2 FPGA-Based IP Watermarking Technique

Field programmable gate array (FPGA) IP generally involves four design level, respectively physical design level, structural design level, behavioral design level and systematic design level. Many IP watermarking techniques are realized at these design level, but IP watermarking techniques at physical design level are the most. Kahng et al. presented to map a watermark into a set of constraints and embedded the watermark using satisfiability (SAT)—a classical NP-complete constraint-satisfaction problem. Yip et al. [11] authenticated a FPGA IP watermark by using public key. Nie et al. [12] proposed a post-layout IP watermarking scheme. The post-layout is abstracted into a graph using graph theory and topology theory. Searching algorithm and optimization algorithm are used in watermark embedding. Khan et al. [13] embedded watermark by rewiring circuit with one or more redundancy addition/removal steps. The watermarked circuit has the same functionality with that of the original after removing these redundancies. If constraints such as timing are satisfied, watermarked circuit could take the place of the original one. But, adding a redundant connection may cause some new redundancies. In order to solve security problem of FPGA based IP design, Wei Liang's team [14–16] proposed several effective and robust methods in watermark embedding and detection. Xu et al. [17] mapped a watermark into positions and some watermark bits (0 and 1). These bits

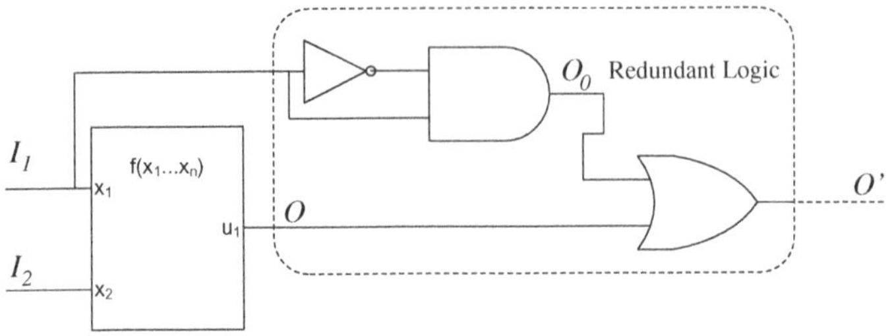

Fig. 2 An example of redundant watermark embedding

are embedded into design in form of redundant logic circuits, as shown in Fig. 2. It causes much less resource overhead. This method can insert more watermark than existing methods due to watermark compression.

At behavioral design level, Raj et al. proposed a watermarking technique for IP identification based on testing in SOC design [18]. It provides high watermark coverage rate, but resistance against collusion attack and hardware overhead require further improvements. Castillo et al. [19] proposed HDL based IP protection scheme by watermarking lookup table (LUT) structure of FPGA. A watermark is inserted between unused LUT and used LUT. However, watermark detection requires adding extra logic. With specific input sequences, this logic will output the watermark data. By comparing to scheme of Raj et al. [18], this scheme is more convenient in watermark extraction. But the added logic is vulnerable to be attacked or removed.

2.1 Secret Key Generation

Most FPGA based IP watermarking utilizes lookup tables (LUTs) structure. Generally, a secret key is required to determine watermark positions. As the key is sensitive information in watermark embedding and extraction, it should be safely reserved. Generally, the generation of secret key requires considering dispersity of positions. It is proper to make the watermarks distribute evenly in the design with high robustness. So, the secret key generation is divided into three steps: resource searching, resource recording and key generation.

Resource searching. An FPGA device always includes configurable logic blocks (CLBs). A CLB includes four slices and there are two LUTs in a slice (e.g. Virtex II FPGA). Firstly, it is necessary to determine the number of unused LUTs in FPGA design. All configurable logic blocks (CLBs) are read in this procedure. Each LUT in CLB is traversed by "Z" shape until all of them are accessed.

Resource recording. During searching procedure, utilization of LUTs in FPGA device is recorded with a two-dimensional table. "0" or "1" respectively denotes a LUT is unused or used.

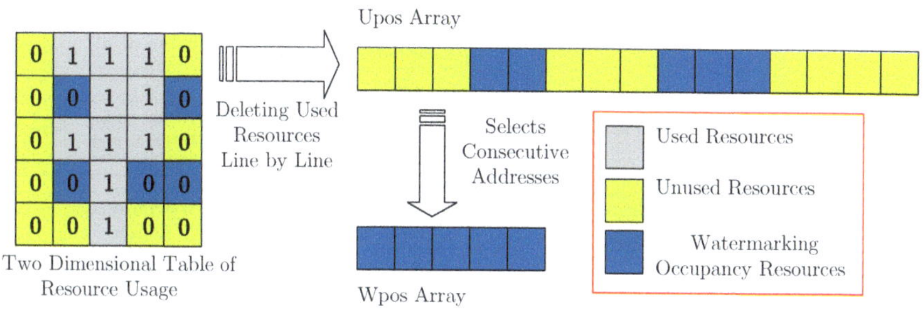

Fig. 3 Secret key generation

Secret key generation. As shown in Fig. 3, recorded utilization of LUTs is recombined as a linear list *Upos*. A block of continuous addresses is selected by random number generator. The linear list stores data of unused LUTs. So, it is highly possible to select continuous addresses that are close to original design. The selected addresses *Wpos* are stored in a key file.

2.2 Watermark Embedding

For FPGA based IPs, the watermarks can be inserted into IP design manually. Namely, some proper positions are searched in physical layout through the design tool (e.g. Xilinx ISE). The watermarks are embedded by configuring the function in caption of the selected positions. Another way utilizes programmable interface provided by device manufacturer. The resource researcher and watermark embedder can be programmed to implement watermarks in bitfile automatically.

A functional soft IP core is described by VHDL language. The design tool (e.g. Xilinx ISE) allocates resources for the IP core. After that, the third-part synthesis or simulation tools (e.g. Synplify and ModelSim) are utilized to map the IP and simulate its functionality. Finally, the physical layout is generated. Constraints in this design, such as timing and area should be set to optimize the design. The watermark positions of LUTs are easily located with the secret key. The watermarks are then inserted into these LUTs by configuring specific function. Besides, some redundant connections are added to hide real watermark positions. Figure 4 shows procedure of FPGA based IP watermark embedding and illustrates some critical steps.

2.3 IP Watermark Extraction

A watermarked IP design may be misappropriated in semiconductor market or illegally used in some products by adversaries. IP owner can apply for a neutral third

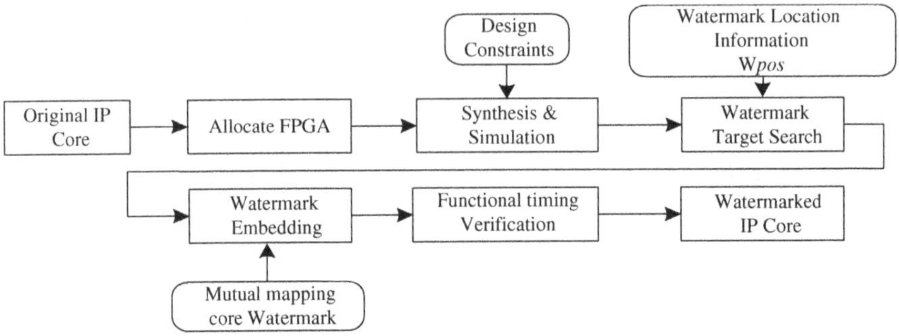

Fig. 4 Procedure of FPGA based IP watermark embedding

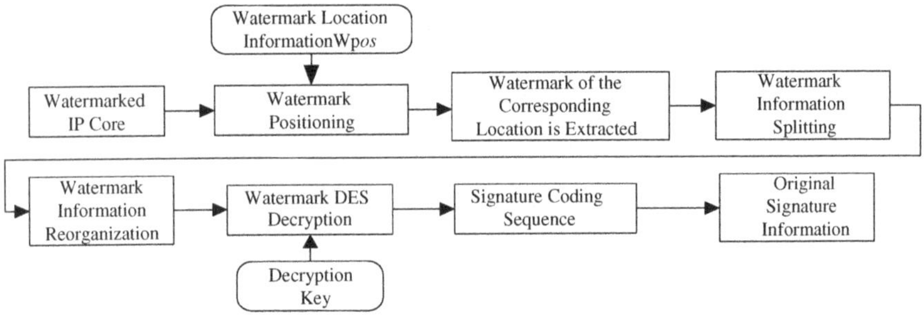

Fig. 5 Procedure of FPGA based IP watermark extraction

party to authenticate the suspected IP ownership. He submits the secret key to the third-party institute and conducts watermark extraction. If a declared watermark is extracted from the suspected IP, the ownership is proven.

Watermark extraction includes watermark locating, splitting, processing, as shown in Fig. 5.

Watermark locating. Generally, IP core is delivered at a low design level (e.g. physical layout) since the use of IPs at physical level is more convenient and easier. So, watermark extraction will locate the watermark positions in *Wpos* and read LUT contents in these positions.

Watermark splitting. The extracted sequence contains encrypted copyright information and mutual mapping factor to reconfigure watermark. With the reserved width, we split the sequence into several parts to reconfigure the original watermark.

Watermark processing. Original watermark is encrypted for better security. With encryption key, the encrypted watermarks in above step can be decrypted. The extracted watermark is compared to the declared watermark for verification. If two watermarks are consistent, the ownership is proven.

Table 1 IP watermarking performance indexes in resource utilization and growth

IP	Device	Length of W	Non-watermarked design			Watermarked design			Growth rate
			L	*L-Num*	$\Delta L(\%)$	*L*	*L-Num*	$\Delta L(\%)$	$\Delta S(\%)$
DES	Xc2v1000	32	3376	10238	32.98	3367	10240	33.04	0.266
STROM	Xc2v1500	64	7308	15357	47.32	7382	15360	47.68	0.272
CACHE	Xc2v2000	128	13234	21521	61.46	13236	21504	61.57	0.305
RS	Xc2v4000	256	25956	46089	56.38	26024	46080	56.44	0.304

2.4 Experiments and Analysis

In this section, we will evaluate and analyze the proposed watermarking algorithm in terms of resource overhead and ability against attacks.

2.4.1 Resource Overhead

In watermark embedding procedure, original watermark information is encrypted by DES algorithm and then hashed. The data can be compressed by using Hash algorithm. Consequently, despite the length of original watermark, the hashed result is 128 bit constantly. The resource overhead will not increase when the length of watermark bits is greater than 128.

Table 1 records some performance indexes in resource utilization and growth. *W* denotes embedded watermark. *L* is the total number of utilized LUTs. *L-Num* represents the total number of LUTs in FPGA device. $\Delta L(\%)$ denotes the rate of utilized LUTs and $\Delta S(\%)$ is the growth rate of utilized resource after watermark insertion. The growth rate of utilized resource is constantly close to 0.3% after embedding watermark, which satisfies the requirements of resource overhead. Since the proposed algorithm utilizes unused LUTs for watermark insertion, the watermark will cause resource overhead. However, the watermarked resources will not be accessed when the system is running. Therefore, the power overhead will not increase. The experiments show that the proposed algorithm has good performance on resource overhead and power consumption.

To evaluate the features of low overhead and high watermark volume, we analyze the resource distribution in original design and watermarked design. Xilinx Virtex II XC2V2000 FPGA device is used in experiments. The RS IP core is selected as the target IP design. Figure 6 shows the resource distribution. The proposed model can improve the number of embedded watermark bits. The rate of resource utilization can be also calculated. Meanwhile, we analyze the resource variation and the resource aggregation is better.

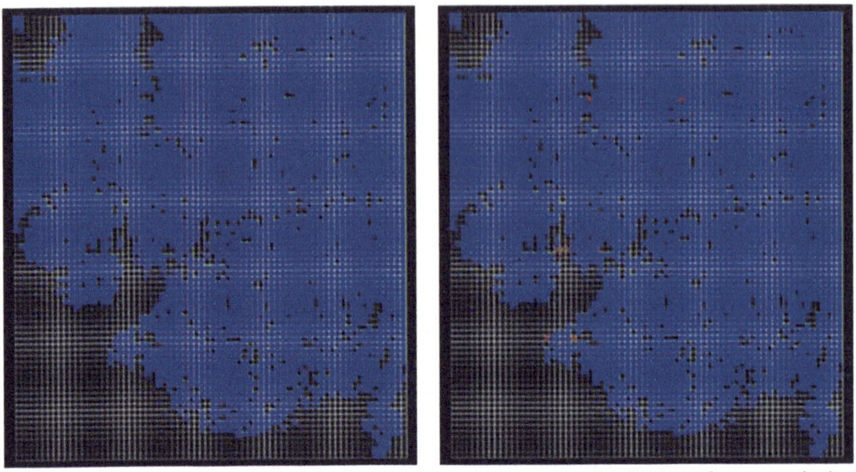

(a) Resource distribution of original RS IP design

(b) Resource distribution of watermarked RS IP design

Fig. 6 Resource distribution of RS IP designs

2.4.2 Security Analysis

The security of IP core mainly reflects the ability of watermark withstanding malicious tampering or attacks. The normal attack methods include removal attack, physical attack, forgery attack and collusion attack etc. The removal attack removes the watermark directly by certain means. For the brute force attack, it searches the inserted secret information by force. The forgery attack inserts the illegal watermark to IP core which should not exist originally. The passive aggression represents that the attacker who can detect the watermark and recognize every mark, but fails to decipher the mark code. The security and performance analysis of proposed algorithm in this paper is conducted under the illegal removal attack and noise attack modes.

Ability against Reverse Analysis Attacks. In the proposed scheme, the watermarks insertion can be implemented by configuration of logic function. It is difficult for illegal attackers to get logic function in programmable logic circuit by reverse analysis attacks. To perform reverse analysis attacks, they should firstly obtain all configuration data of FPGA design. There are two ways to get configuration data generally. One is to steal the bit stream and another one is to read configuration data in RAM by using micro-probe.

With the way of stealing bit stream, attackers need to import the programmable data in every time of system booting. The way makes it possible to analyze circuit function from bit stream. In our proposed scheme, a stabilized power is used to keep the information in storage nonvolatile. The configuration data is no need to be imported again in system booting. In this case, the attackers cannot steal the bit stream of IP circuit.

Besides the way of stealing bit stream, attackers may use micro-probe to read configuration information in RAM. Therefore, the RAM units and the output signal in our scheme are set at the low level of chip. The attackers cannot probe related configuration logic by micro-probe. Consequently, IP circuits with our proposed watermark scheme has good ability against reverse analysis through stealing bit stream, especially reverse analysis on layout.

The noises in above experiments are Gaussian noise. In following experiments, we focus on noise attacks of GGD type and MSS type. The noise intensity is denoted by P, 0<P<1. Figure 7c compares the proposed scheme with the method based on one dimensional chaotic mapping (ODCM). The experimental results in Fig. 6c show that the performance of ODCM against GGD noise attack is low with the increase of P. The reason is that the position aggregation parameter becomes small after suffering GGD noise attacks when P increases. In this case, the error probability of IP circuit increases correspondingly. In Fig. 7d, when P becomes larger, our scheme has better ability against MSS noise attack by comparing with that in literature [20].

Noise Attacks. The signals of the watermarked circuits with our scheme are not in Gaussial distribution. Where ξ denotes the optimal threshold for attack of noises. Using optimization methods in [21] when $\xi = 0.2, 0.4, 0.6, 0.8$, we compare the

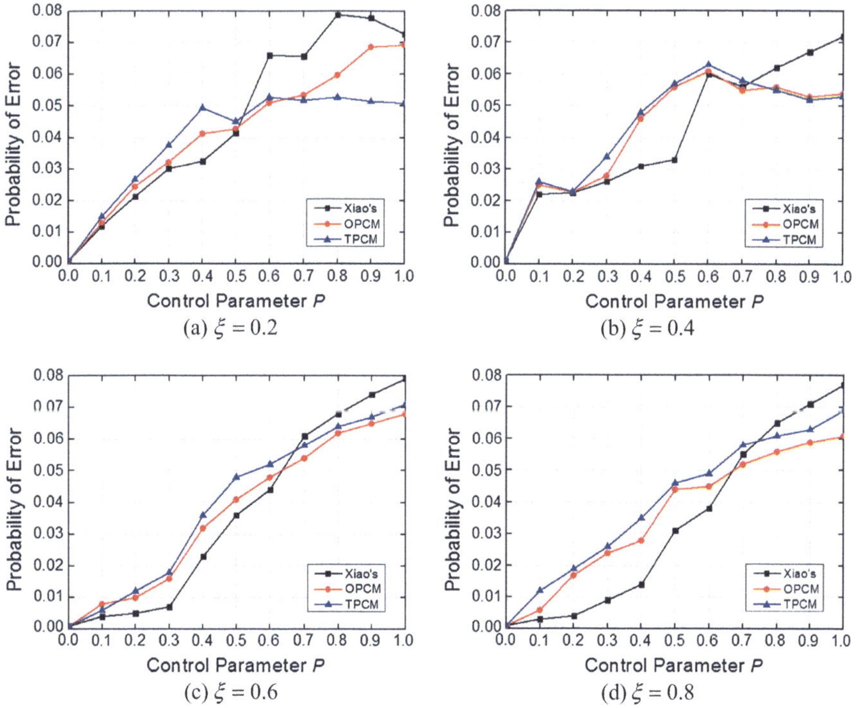

Fig. 7 When $\xi = 0.2, 0.4, 0.6, 0.8$, the performance of various algorithms in terms of resistance to noise attack

performance of resistance against noise attack. The performance after suffering noise attacks can be obtained by using numerical method. Figure 7a shows a comparison of OPCM [14] and TDCM scheme with that in literature [20]. With $P < 0.6$ and low noise intensity, the security of two schemes are better than that in [20]. Figure 7b shows the ability against noise attacks of our proposed two schemes are better than the method based on one dimensional chaotic mapping when $P > 0.9$. In contrast, the proposed method is superior to previously proposed approaches in terms of resistance against noise attack.

3 FSM-Based IP Watermarking Technique

Finite state machine (FSM) based IP watermarking has also been widely studied. Torunoglu et al. [22] utilized unused transitions in state transition graph (STG) for watermarking. As shown in Fig. 8, some new transitions are added in original STG. The watermark is indicated by creating a Euler path. Oliveira et al. [23] divided a 128-bit watermark into a set of bit fragments, as input sequence. A designer modifies state in STG to insert watermark. To enhance the detectability of FSM-based IP watermark, Abdel-Hamid et al. [24] added watermarks into FSM of sequential circuit. This scheme generates various transition adding solutions under control of different key. With initial state and input sequence of watermark, it is convenient to detect watermark from output sequence. Cui et al. [25] proposed an adaptive watermarking technique by modeling some closed cones in an originally optimized logic network (master design) for technology mapping. IP watermark in this scheme achieves low overhead and good resistance against attacks.

For complex logic circuit, STG is implemented by modifying some components of circuit. Different with traditional modification of STG, addition of delay state will not affect output value. If a watermark is implemented in this way, watermark removal will be complex and time-consuming. It makes an alteration to state coding. When value of state variable is not a watermark, new value of state variable will be changed accordingly. By adding two transcoders, newly generated state variable will

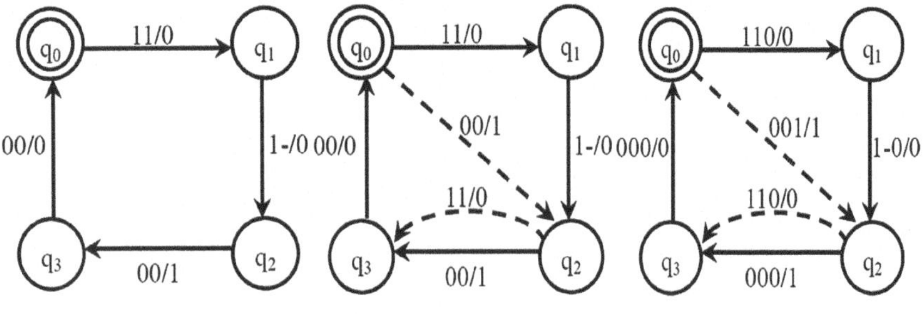

Fig. 8 An example of FSM based IP watermarking

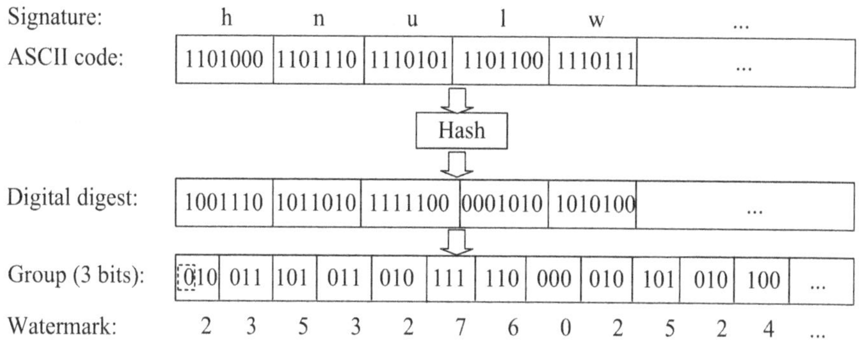

Fig. 9 Watermark generation

change the output. Besides, its value changes as well for any input except $a_1, ..., a_m$. The transcoder is realized by a series of linear transformation. A variable set $X = \{x^1, ..., x^i, ..., x^j, ..., x^n\}$ is transformed as $X\prime = \{x^1, ..., x^i, ..., x^i \oplus x^j, ..., x^n\}$. Any function $F : X \rightarrow \{0, 1\}$ is mapped to $F\prime : X \rightarrow \{0, 1\}^k$. A series of elementary transformations finally realizes any linear transformation by adding two EXOR gates and a gate in transcoder.

Consequently, a FSM-based watermarking scheme is proposed to protect reused IP. When IP dispute occurs, IP owner extracts the maximal delay state set through state transformation relations among circuit signals. Finally, it proves IP ownership.

3.1 *Watermark Generation*

Since only binary signals can be traced in IP design, a signature should be transformed into a binary sequence. The generated sequence will be then disordered by using a hash function. The digest is divided by three bits (left zero padding) and each group denotes a decimal number (0–7). Let a signature be "hnulw…" and watermark generation process is shown in Fig. 9. "0" in dotted rectangle is left zero padding.

3.2 *Watermark Embedding*

In this section, we introduce watermark insertion by modifying state delay information in STG, as described in follow steps.

Input: a watermark W and an IP core

Output: watermarked IP core

Step 1: Traverse each state $s_i \in S$ of STG with a sequence of inputs $a_1, ..., a_m$ and collect a set of state transitions $R\ (T)$.

Step2: Analyze all the state delay information in delay states $R(T)$ and set an appropriate threshold T_N as criteria for selecting $R'(T)$. $R'(T)$ includes states suitable for modification. Selection of T_N depends on the type of an IP core.

Step3: Randomly select γ state delay values from $R(T)$ by considering length of a watermark and create a set of delay states $R'(T)$ for watermark insertion.

Step4: Analyze delay state values in $R'(T)$ at specific positions. Replace the last number of each delay state value with a watermark fragment. This operation is repeated until all watermark fragments are inserted. In this case, a watermarked IP design is generated finally.

3.3 Watermark Extraction

When an IP dispute occurs, IP owner can apply to authenticate the ownership by extracting watermarks from the suspected IP. The watermark is embedded into STG of IP design. The concrete extraction procedure is illustrated as follows.

Input: a watermarked IP core

Output: digest of watermark W

Step1: Extract and analyze STG of the watermarked IP core.

Step2: Traverse each state $s_i \in S$ in STG with a sequence of inputs $a_1, \ldots, a_m$ and obtain a set of state transitions, denoted by $R\ (T)$.

Step3: Obtain a set of delay states $R(T)\prime$ and analyze the watermarked STG.

Step4: Extract γ watermarked state delay information with random selection rule used in watermark embedding. The last number can be extracted by analyzing state delay information and transformed into binary sequence

Step5: Recombine the binary sequence through the reverse procedure of embedding and finally get the embedded watermark digest.

Verification is implemented by comparing the extracted digest to the declared one.

3.4 Experimental Results

The proposed method has been tested on Xilinx VirtexII device XCV600 by watermarking three public cores with 128 bits watermark: DES56, ALU, RSA. The performances in the form of timing, SNR and resources are primarily verified. The test results are shown in Table 2.

Table 2 reveals that DES core utilizes the most CLBs, while ALU the least for the three cores. The core with the maximal delay is DES occupied the most resources, followed by RSA, ALU. By comparison with methods in [26, 27], the proposed method is not the best in terms of timing performance. While the SNR and the occupied resource relative to original circuit are both lower. Therefore, the proposed method has lower impact on circuit function, better security and resource overhead.

Table 2 Performance comparison of different IP core physical layouts

Method	Core	Devices	Used slices	Timing(ns)	SNR	% Resources
[26]	DES	XCV600	972	7.706	0.432	0.786
	RSA	XCV600	668	9.103	0.503	1.899
	ALU	XCV600	481	15.122	0.422	2.591
[27]	DES	XCV600	958	8.416	0.716	0.558
	RSA	XCV600	683	9.706	0.706	2.793
	ALU	XCV600	485	16.231	0.231	2.883
Our method	DES	XCV600	947	7.802	0.602	0.367
	RSA	XCV600	656	9.901	0.491	1.707
	ALU	XCV600	479	17.998	0.368	2.165

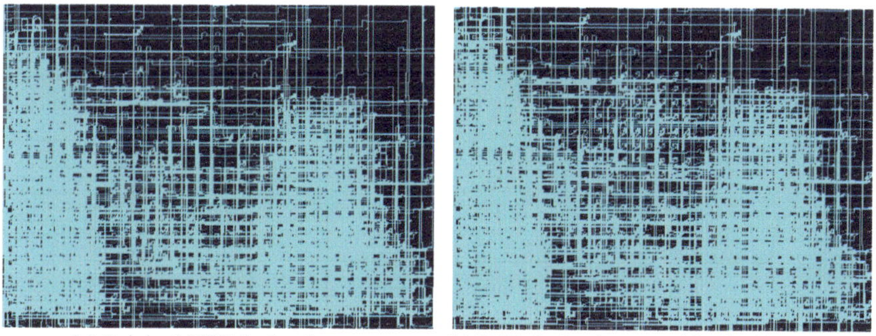

(a) Original DES design layout (b) DES design layout with 128 bitwatermark

Fig. 10 Original DES design layout and the layout with 128 bitwatermark

Figure 10 shows the experimental results for DES core. The physical layouts reveal that, the watermarked layout in Fig. 10(b) has higher density of occupied resource, but lower impact on circuit function in comparison with the original in Fig. 10(a).

4 DFT-Based IP Watermarking Technique

Digital watermarking applied in design-for test (DFT) has been extensively concerned. Most of the DFT watermarking techniques focus on scan chains. In the methods proposed by Fan et al. [28], the watermark generation is integrated in the test module. Five possible methods for watermark hiding are presented. Since the test circuit instead of the IP core is marked independently, it is vulnerable to removal attacks. Saha et al. [29] proposed to watermark both the scan tree and single scan chain, separately embedding the signatures of the owner of physical design tool and that of the logic design tool. Cui et al. [30] proposed to insert watermark through

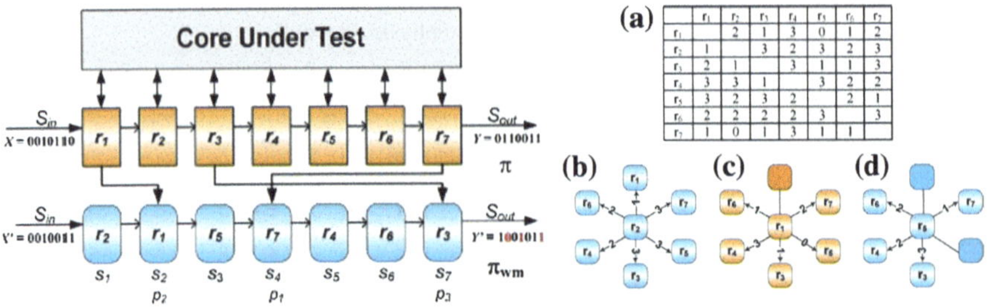

Fig. 11 IP watermark implementation by reordering scan cells

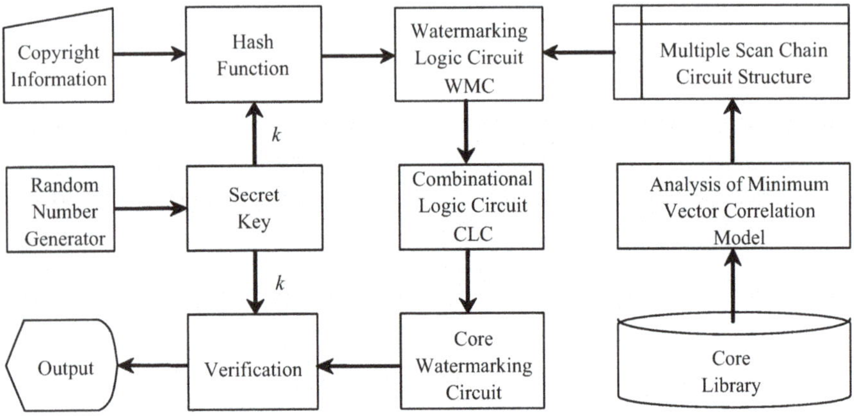

Fig. 12 Overview of multiple scan chains based IP watermarking scheme

reordering the scan cells in a single scan chain minimizing power overhead, as shown in Fig. 11.

In this section, we introduce an IP protection method by watermarking multiple scan chains in sequential circuit. This scheme adopts DFT test model in SOC design, and uses an LFSR for pseudo random test vector generation. Let the structure of multiple scan chains be M. The multiple scan chains M can be transformed into M_p with the minimum correlation $\Phi\left(M_p\right)$ after exchange operations. M_p is suitable for watermark embedding.

The overview of multiple scan chains based IP watermarking scheme in sequential circuit is described in Fig. 12. The copyright information is encrypted and transformed using hash function with private key k. On basis of the minimum correlation model and multiple scan chains, a watermarking logic circuit (WMC) is designed to change states of specific registers in multiple scan chains for watermarking the design. The watermark can be effectively detected without interference with normal function of the circuit, even after the chip is packaged.

4.1 Watermarking Architecture

A watermarking structure of multiple scan chains with minimum correlation is introduced in this section. Figure 13 shows an example for watermarking multiple scan chains. Assume that, the circuit under test consists of 6 scan cells s_i, $i = 1, 2, \ldots, 6$, these cells are organized into two scan chains $c_1 = \{s_1, s_2, s_3\}$ and $c_2 = \{s_4, s_5, s_6\}$. In the watermark circuit, one input of XOR gate is connected to one cell in multiple scan chains, and another controlled by watermark enable signal *w_en* and output of arbitration logic circuit (ALC). However, output of ALC is under the control of states in LFSR.

4.2 Watermark Embedding

The signature, representing one's identity, is encrypted and then hashed. The generated digital digest is inserted into IP core as watermark. Hash function H is a transformation using x as input and the returned value is called hash value, denoted by h, i.e. $h = H(x)$. Since hash is a one-way function, given a value h, it is computationally impossible to calculate x by using $H(x) = h$.

A signature is hashed by MD5 for a 128-bit digest ξ. In preprocessing procedure, ξ is transformed into binary sequence $< \beta_1, \beta_2, ..., \beta_i ... \beta_n >$. The chaos system generates a key sequence $\kappa_s < \kappa_{s1}, \kappa_{s2}, ..., \kappa_{sj}, ..., \kappa_{sp} >$. The sequence $< \beta_1, \beta_2, ..., \beta_i ... \beta_n >$ is mapped to a set of watermark fragments $\{< \varpi_1, \varpi_2, ..., \varpi_j, ..., \varpi_p > | \varpi_j =< \beta_{k,...r} >\}$. So, $\{\gamma(\varpi_j) | 1 < j < p\}$ is utilized to control register positions as a set of constraints. A scan chain with the minimum correlation $< s_1, s_2, ..., s_i, ..., s_\lambda >$ is selected for watermarking. The arbiter logic circuit limits constraints $\gamma(\varpi_j)$ to positions of specific scan chains. Figure 14 shows an example of multiple scan chains based watermarking scheme.

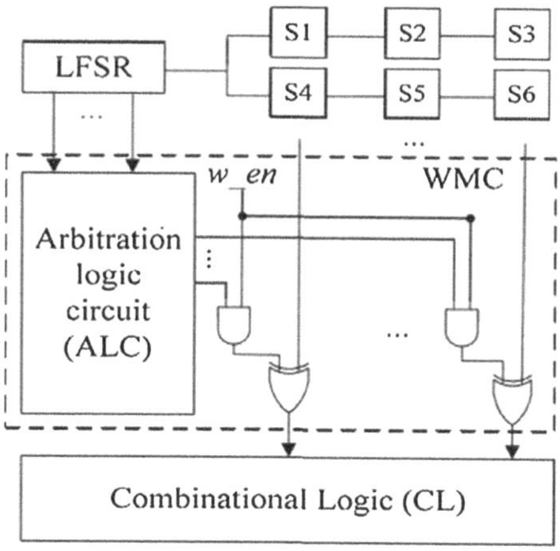

Fig. 13 The watermarking architecture of multiple scan chains

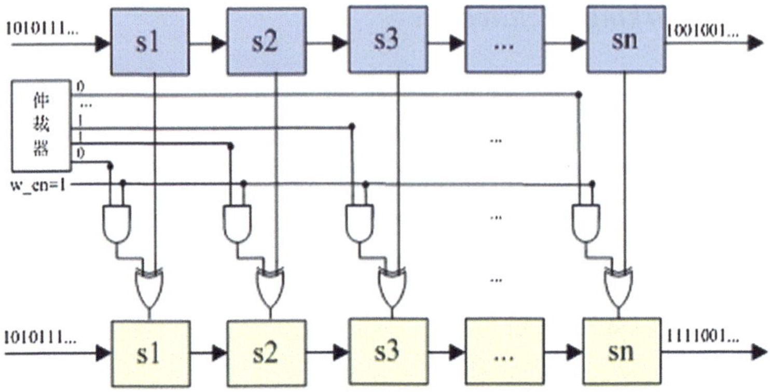

Fig. 14 An example of multiple scan chains based watermarking

There are two modes, normal model and watermark model. In the normal mode ($w\_en = 0$), the circuit under test executes normal scan test and in watermark mode ($w\_en = 1$), a specific state shifted in ALC may cause 1, thus values of some cells in multiple scan chains will be reversed and then be output. The IP identification could be verified by comparing the output in normal mode and watermark mode for the same input vector.

4.3 Watermark Extraction

When the IP core is suspicious to be misappropriation, the author could apply to the third party for the verification of watermark by the following steps.

First of all, we read in the watermarked design and insert architecture of multiple scan chains. LFSR is used for the generation of test vectors. At present, $w\_en = 1$, the watermark circuit is active. The test vectors are shifted in multiple scan chains. The response vectors will be output through the combinational logic in the test circuit. Therefore, the watermarked responses R_m could be detected at the scan output. Then we set w_en signal as '0', now the scan results become the original response R since the watermark circuit is not active. Accordingly, given a specific input vector, by comparing the response vector R and R_m, respectively before and after watermark, the watermark positions will be found. After a series of transformations, the watermark fragments distributed in the whole design are found. Using the stored sequence $Rn(k)$, the watermark fragments can be recombined as an extracted watermark $W_m\prime$. The IP identity could be verified by comparing W_m and $W_m\prime$.

4.4 Experimental Results and Performance Analysis

The proposed scheme by watermarking multiple scan chains with the minimum correlation is implemented in VC on a 1.2 GHz Sun UltraSPARC-T1 machine. The watermarking scheme is applied on sequential circuits from ISCAS'85, ISCAS'89 and ISCAS'99 benchmark suites. The performance analysis of the proposed scheme will focus on resource overhead, resistance to attacks and comparison of experimental results.

4.4.1 Resource Overhead

It is critical to evaluate the resource overhead after watermarking. We select five circuits with the gate number over thousand for experiments. The zero delay models in [31] are used for resource evaluation, through which the transition times will be computed for reflecting the actual resource overhead. The experiment is conducted by the following steps:

(1) Generate the pseudo random vectors by using LFSR and construct the optimal scan architecture with the minimum correlation, and then output the test vectors;

(2) Load the test vectors in the circuit under test and record the transitions of internal nodes, and then calculate peak power and average power;

(3) Partition the test point in sequential circuit according to the architecture of multiple scan chains;

(4) Use LFSR for the generation of test vectors once again, and obtain the watermarked response vectors; calculate the peak power and average power after watermark by recording the transitions of internal nodes during test.

As shown in Table 3, the cells number of combinational circuit and sequential circuit are shown in column 2 and 3 respectively. The columns, "P_w", "P_f" and "ΔK" are respectively the average power, peak power and the coverage rate of the test nodes. The experimental results in Table 3 show: the average power and peak power both reduce accordingly, while the coverage rate increase slightly. It proves that the proposed scheme has the advantages of lower resource overhead and higher coverage rate without affecting the normal circuit function.

4.4.2 Comparison of Experiments

The experiments are conducted on the multiple scan chains with the minimum correlation. The comparison results of the proposed scheme with methods in [32, 33] are shown in Table 4.

Assume that, $\Phi(M_p)$ is the minimum correlation of scan chains, P_c denotes the probability of coincidence and ΔS represents the coverage rate of watermark detection. Table 4 shows that the proposed scheme has lower P_c than other methods, which verify the stronger resistance of our scheme to attacks. The coverage rate of water-

Table 3 The performance comparison of the original and watermarked circuit

Circuit	Combinational Logic N	Sequential Logic C	Original Circuit			Watermarked Circuit		
			P_w	P_f	ΔK(%)	P_w	P_f	ΔK(%)
S5378	2779	179	3797	1968	84.56	3461	1876	90.11
S9234	5597	211	6785	3622	90.12	6123	3601	98.54
S13207	7952	669	10908	6471	82.16	9875	5947	88.34
S35932	16066	1728	41235	19677	89.64	32471	17983	91.44
S38584	19354	1546	20657	14906	92.82	18195	12876	96.01

Table 4 Comparison of watermarking methods

Circuit	$\Phi(M_p)$	Proposed		[33]		[32]	
		P_c	$\Delta S(\%)$	P_c	$\Delta S(\%)$	P_c	$\Delta S(\%)$
i7	18	2.17E-21	91.02	2.91E-21	86.49	7.52E-20	90.23
i9	19	1.02E-14	90.62	3.49E-14	85.07	1.05E-13	88.48
i2	22	1.91E-23	97.83	6.38E-23	83.24	2.64E-19	79.41
i8	15	5.77e-32	94.27	1.67E-32	88.36	2.60E-31	91.86
frg2	14	1.23E-19	92.06	6.02E-18	91.68	1.91E-19	70.77
alu4	25	1.93E-41	94.82	7.14E-34	79.91	1.70E-39	86.09
apex6	20	5.77E-33	99.15	3.06E-24	95.28	8.16E-31	93.62
rot	20	4.98E-26	100.00	8.75E-25	94.76	1.41E-21	87.71
x3	18	4.66E-36	95.37	8.08E-25	89.41	6.28E-35	71.44
k2	33	2.42E-32	96.53	3.24E-32	92.09	8.64E-32	83.57

mark detection ΔS s larger. Due to the architecture of multiple scan chains we use in the scheme, the watermark has become more observable and testable. Therefore, the proposed scheme has lower probability of coincidence P_c and better coverage rate of watermark detection.

5 Self-recoverable Dual IP Watermarking Technique

Robustness is an important metric of IP watermarking technique. However, majority of existing methods cannot recover impaired watermarks after suffered from attacks, causing a failure of ownership authentication. In this section, we introduce an FPGA based dual IP watermarking technique with ability of self-recovery. It authenticates IP ownership even watermark is suffered from illegal attacks and causes lower watermark embedding overhead.

5.1 Watermark Generation

IP circuits has two signals "0" and "1". So, ownership information is firstly transformed into contents that are suitable for circuit. This section generates dual IP watermarks, respectively denoted by binary sequences $s = s_0 s_1 s_2 \ldots s_n$ and $s\prime = s\prime_0 s\prime_1 s\prime_2 \ldots s\prime_n$. A watermark indicates ownership information (signature) of IP owner and another watermark represents identity of IP user. In this case, dual IP watermarks can authenticate IP ownership and monitor the use of IPs.

5.2 Watermark Embedding

Generally, the constraints in bitfile should be modified to limit location of watermarked LUTs close to the functional LUTs. It avoids high resource occupation and delay caused by long connections after inserting watermark. The detailed process includes following steps.

(1) Breadth first search and depth first search methods are utilized to locate slices in CLBs. For Virtex II FPGA, there are two LUTs in a slice, F and G. Whether a LUT in a slice is used can be determined by values of F and G in LUT. The values "0" and "1" respectively indicate unused and used. The coordinates of unused LUTs are recorded for selection of watermark positions.

(2) The dual watermarks $s = s_0 s_1 s_2 \ldots s_n$ and $s = s\prime_0 s\prime_1 s\prime_2 \ldots s\prime_n$ are divided by 16 bits. Each group relates with a coordinate of LUT. So, an index table δ is created. Here s is the primary watermark and $s\prime$ is the secondary watermark;

(3) An ordered pair (k, m) satisfying $k \leq m$ is selected to create a polynomial $f_1(x) = a_0 + a_1 x + \ldots + a_{k-1} x_{k-1}$. Here the value of x can be $0, 1, 2, \ldots, m$ and $2 \leq k \leq m$. $a_0, a_1, a_2, \ldots, a_{k-1}$ is a sequence of randomly selected coefficients, $a_0 = s$, $H_k = f_1(i_k)$, $i_k \in [0, m]$. In this case, the reconfiguration information of one signature is computed, denoted by $H = \{H_k | k = 1, 2, \ldots, m\}$. By the same way, we get the reconfiguration information of another signature, denoted by $H\prime = \{H\prime k | k = 1, 2, \ldots, m\}$. The H and $H\prime$ are reserved as parameters in watermark recovery.

(4) Select four hexadecimal numbers from one signature $s = s_0 s_1 s_2 \ldots s_n$. The reconfiguration information of both signatures is decomposed into $A \times B + C$. C denotes the information being inserted in location (A, B). The insertion procedure is then performed, namely changing the value at corresponding position in self-constraint file of bitfile. For better security, embedded bits will be encrypted with the private key *key*$\prime$. The reconfigurable information corresponding to *key*$\prime$ is $H\prime$. With the same steps, the secondary watermarks can be also processed. Here the embedded bits are encrypted with private key *key*. The reconfigurable information corresponding to key is H.

(5) Each LUT implements 16 bits' watermark by configuring specific logic functions. Watermark embedding is finished until all watermark bits are inserted in redundant attribute identifiers.

5.3 Watermark Extraction

To authenticate the ownership of an IP, the embedded watermark should be also extracted in this scheme. The extracted watermark will compare to the declared one. If they are consistent, the ownership can be successfully authenticated. But if the extracted watermark has some errors, the task of watermark recovery will be activated. Dual IP watermark extraction includes following steps.

(1) Extract redundant attribute identifiers. If the watermarks are not impaired, we can find all newly inserted redundant attribute identifiers in self-constraint file of bitfile with private keys *key*, *key'*, and the reserved watermark locations.

(2) Reconfigure index table and mapping relation of redundant combinational expression. With position parameter μ of embedded redundant attribute identifiers, the index table and redundant combinational expression $A \times B + C$ can be computed. Thus, we can get the positions of LUTs corresponding to the value of C in index table.

(3) With the inverse process of watermark transformation, we can extract redundant attribute identifier in index table and compute related logic expression. The extracted information is transformed to get hashed value. If it matches that of original signature, the IP is authentic.

5.4 Watermark Recovery

In traditional IP watermarking techniques, watermarks are difficult to recover if being damaged by adversary. The ownership cannot be authenticated with an impaired watermark. To address this issue, a watermark recovery scheme is proposed to authenticate ownership after being suffered from attacks. It depends on the thought of key reconfiguration in secret sharing mechanism. When IP dispute occurs, IP owner can extract and recover impaired watermarks C_1 and C_2. Dual IP watermarks $s = s_0 s_1 s_2 ... s_n$ and $s\prime = s_0 s_1 s_2 ... s_n$ are mutually relevant. There are two cases in watermark recovery. (1) If a part of watermark C_1 is damaged, it can be recovered by $s\prime$, namely $C_1 = E^{-1}(F(x_2), \rho)$. $F(x_2)$ is the main mapping function of $s\prime$. ρ denotes self-recovery factor. (2) If another part of watermark C_2 is damaged, s could recover C_2 by calculating $C_2 = E^{-1}(F(x_1), \rho)$. $F(x_1)$ is the main mapping function of s.

When IP watermark is impaired after suffering from attacks, watermark recovery can be used to extract correct signature for successful IP authentication. The flow of dual IP watermark recovery is shown in Fig. 15. Relevancy stream P is extracted from watermark stream S' and encoded as $P' = \{f(x_i) | i = 1, 2, ..., k\}$. Finally, P' is utilized as sub-key for reconstructing original signature. Watermark M' can be restored by reconfiguring $f(x)$ and transformed into original watermark finally.

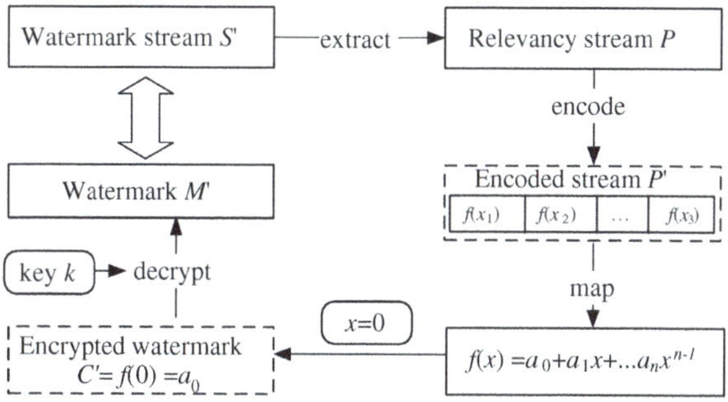

Fig. 15 Flow of dual IP watermark recovery

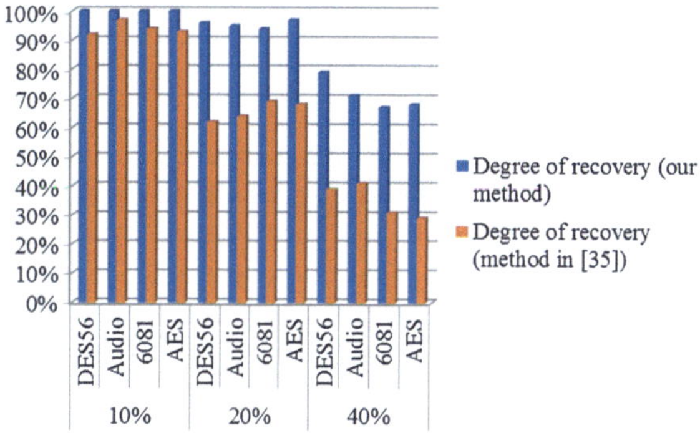

Fig. 16 Evaluation and comparison of watermark recovery

5.5 *Performance Evaluation*

We conduct experiment to evaluate the resistance against removal attack. The length of embedded watermarks is 512 bits. The results with impaired watermarks of 10, 20 and 40% are compared to method in [34]. The comparison is shown in Fig. 16.

After suffering from removal attack, successful recovery of 70% watermarks is regarded as criterion of acceptability. In Fig. 16, with the increase of impaired watermarks, watermark recovery leads to increase of resource and path delay. But if there are 20% impaired watermarks, method in [34] cannot achieve the recovery criterion. The more embedded watermarks are, the more occupation of LUTs is. If impaired watermarks reach 40%, the proposed method has a high percentage to recover impaired watermarks. But method in [34] cannot realize watermark recovery in this case. So, the resistance against removal attack is encourage in the proposed method.

6 Conclusion

IC chip is the basic equipment in IOT environment. IP reuse technique brings convenience, but also cause risk of copyright being infringed. Many watermarking schemes are proposed to address IP protection problems. Reasonable IP watermark embedding and extraction scheme provide protection at various design levels of IP designs. This chapter introduces several types of IP watermarking techniques. It is focused on the intellectual protection problem of the very large integration circuit and a novel algorithm which is suitable for the IP protection of integration circuit has been proposed. These techniques realize improvements on previous work and have great significance to protect reused IPs in IC designs. They succeeded in reducing the power consuming as well as largely increasing the watermark information concealment of the safety modal. Thus, it indeed improved the resistance ability of the watermark algorithm against the illegal attacks

Although the intellectual property core watermark technique has provided many effective watermark algorithms for the research area of integration circuit secure design in recent years, these achievements are still far away from maturation in industrial application. Thus, more research and exploration is still required to find the solution which has a high recognition by both academic and industrial fields.

Acknowledgements This work is supported by the National Science Foundation of China (61572188), the Research Project supported by Xiamen University of Technology (YKJ15019R, YSK15003R), Xiamen Science and Technology Foundation (3502Z20173035).

References

1. Koushanfar F, Fazzari S, McCants C, et al. 2012. Can EDA combat the rise of electronic counterfeiting? In *Proceedings of 2012 49th ACM/EDAC/IEEE design automation conference (DAC)*, 133–138.
2. Majzoobi M, Koushanfar F, Devadas S. 2010. FPGA PUF using Programmable Delay Lines. In *Proceedings of information forensics and security (WIFS)*, 51–65.
3. Guajardo J, Guneysu T, Kumar S S, et al. 2009. Secure IP-block distribution for hardware devices. In *IEEE international workshop on hardware-oriented security and trust*, 82–89.
4. Kirovski D, Potkonjak M. Local watermarks: Methodology and application to behavioral synthesis. *IEEE Transactions on Computer-Aided Design of Integrated Circuits and Systems* 1277–1283.
5. Marsh C, Kean T. 2007. A security tagging scheme for ASIC designs and intellectual property cores. *Design & Reuse*, 57–64.
6. Goren S, Ugurdag H F, Yildiz A ,Ozkurt O. 2010. FPGA design security with time division multiplexed PUFs. In *Proceedings of international conference on high performance computing and simulation (HPCS)*, 608–614.
7. Lach J, Mangione W H, Potkonjak M. 2001. Fingerprinting techniques for field-programmable gate array intellectual property protection. *IEEE transactions on computer-aided design of integrated circuits and systems*, 1253–1261
8. Guneysu T, Moller B, Paar C. 2007. Dynamic intellectual property protection for reconfigurable devices. In *Proceedings of the 15th annual IEEE symposium on FPT*, 287–288

9. Li, D., W. Zheng, and M. Zhang. 2007. Development of IP watermarking techniques. *Journal of Circuit and Systems* 12(4): 84–92.
10. Roy J A, Koushanfar F, Markov I L. 2008. EPIC: Ending piracy of integrated circuits. In *Proceedings of the conference on design, Europe*, 1069–1074.
11. Yip K, Ng T. 2000. Partial-encryption technique for intellectual property protection of FPGA-based products. *IEEE Transactions on Consumer Electronics*, 183–190.
12. Nie T, Liu H, Zhou L. 2012. A time-constrained watermarking technique on FPGA. In *Proceedings of 2012 international conference on industrial control and electronics engineering (ICICEE)*, 795–798.
13. Khan M and Tragoudas S. 2005. Rewiring for watermarking digital circuit netlists. *IEEE transactions on computer-aided design of integrated circuits and systems*, 1132–1137.
14. Liang, W., X. Sun, Z. Xia, and J. Long. 2011. A chaotic IP watermarking in physical layout level based on FPGA. *Radioengineering* 20(1): 118–125.
15. Liang, W., K. Wu, H. Zhou, and Y. Xie. 2015. TDCM: An IP watermarking algorithm based on two-dimensional chaotic mapping. *Computer Science and Information Systems* 12(2): 823–841.
16. Liang W, Long J, Chen X, Xiao W. 2016. Publicly verifiable blind detection for intellectual property watermarks through zero-knowledge protocol. *International Journal of System Assurance Engineering and Management*, 738–981.
17. Xu J B, Long J, Liang W. 2011. A DFA-based distributed IP watermarking method using data compression technique. *Journal of Convergence Information Technology*, 152–160.
18. Raj N, Josprakash, et al. 2011. Behavioral level watermarking techniques for IP identification based on testing in SOC design. In *Proceedings of international conference on information technology and mobile communication*, 485–488.
19. Castillo E, Meyer-Baese U, García A. 2007. IPP@HDL: Efficient intellectual property protection scheme for IP cores. *IEEE Transactions on VLSI Systems*, 578–591.
20. Sun, X., M. Zhang, and H. Zhang. 2013. *Two-Dimension Chaotic-Multivariate Signature System* 10(1). 1694–0814.
21. Basu, A., D.B. Roy, and D. Banerjee. 2011. FPGA implementation of IP protection through visual information hiding. *International Journal of Engineering Science and Technology* 3(5): 4191–4199.
22. Torunoglu I, Charbon E. 2000. Watermarking-based copyright protection of sequential functions. *IEEE Journal of Solid-State Circuits*, 434–440.
23. Oliveira A L. 2001. Techniques for the creation of digital watermarks in sequential circuit designs. *IEEE Transactions on Computer-Aided Design of Integrated Circuits and Systems*, 1101–1117.
24. Abdel-Hamid A T, Tahar S. 2008. Fragile IP watermarking techniques. In *Proceedings of NASA/ESA conference on adaptive hardware and systems*. Noordwijk, 513–519.
25. Cui A, Chang C H, Tahar S. 2008. IP watermarking using incremental technology mapping at logic synthesis level. *IEEE Transactions on Computer-Aided Design of Integrated Circuits and Systems*, 1565–1570.
26. Yuan L and Qu G. 2004. Information hiding in finite state machine. In *Information hiding workshop*, 340–354.
27. Abdel-Hamid A T, Tahar S, and Aboulhamid E M. 2006. Finite state machine IP watermarking: A tutorial. In *Proceedings of the first NASA/ESA conference on adaptive hardware and systems (AHS'06)*, 457–464.
28. Fan Y. 2008. Testing-based watermarking techniques for intellectual-property identification in SOC design. *IEEE Transactions on Instrumentation and Measurement*, 467–479.
29. Saha D, Sur-Kolay S. 2010. A unified approach for IP protection across design phases in a packaged chip. In *Proceedings of 23rd international conference on VLSI design*, 105–110.
30. Cui A, Chang C H. 2012. A post-processing scan-chain watermarking scheme for VLSI intellectual property protection. In *Proceedings of 2012 IEEE Asia pacific conference on circuits and systems (APCCAS)*, 412–415.

31. Khan, M., and S. Tragoudas. 2005. Rewiring for watermarking digital circuit netlists. *IEEE Transactions on Computer-Aided Design of Integrated Circuits and Systems* 24(7): 1132–1137.
32. Cui, A., Chang, C. H. 2008. Intellectual property authentication by watermarking scan chain in design-for-testability flow. In *Proceedings of International Symposium on CAS*, 2645–2648.
33. Kirovski, D., Y.Y. Hwang, et al. 2006. Protecting combinational logic synthesis solutions. *IEEE Transactions on Computer-Aided Design of Integrated Circuits and Systems* 25(12): 2687–2696.
34. Xu, J., Y. Sheng, W. Liang, L. Peng, and J. Long. 2016. A high polymeric mutual mapping IP watermarking algorithm for FPGA design. *Journal of Computational and Theoretical Nanoscience* 13(1): 186–193.

Cyber Defence Capabilities in Complex Networks

Dragoş Ionică, Nirvana Popescu, Decebal Popescu and Florin Pop

Abstract This chapter presents a quick overview about the existing cyber Defence capabilities and cyber ranges in complex networks where operations testbeds meant to bring improvements in cyber security training. The current chapter gives a brief on the problem area where cyber developments within the Ministry of Defence (MoD) are introduced along with the test range. The research goal is introduced as well as the study limitations and desired results. The chapter ends with some recommendations and suggestions that the researcher came up with based on the results of the study for complex networks.

1 Introduction

This chapter presents a quick overview about the existing Cyber Ranges and the computer network operations testbeds meant to bring improvements in cyber security training. The current introductory chapter gives a brief on the problem area where cyber developments within the Ministry of Defence (MoD) are introduced along with the test range. The research goal is introduced as well as the study limitations and desired results. Finally, a general idea about the research methodology is presented.

D. Ionică · N. Popescu · D. Popescu · F. Pop (✉)
Computer Science Department, Faculty of Automatic Control and Computers, University Politechnica of Bucharest, Bucharest, Romania
e-mail: florin.pop@cs.pub.ro; florin.pop@ici.ro

D. Ionică
e-mail: ionicadrgs@gmail.com

N. Popescu
e-mail: nirvana.popescu@cs.pub.ro

D. Popescu
e-mail: decebal.popescu@cs.pub.ro; decebal.popescu@ici.ro

D. Popescu · F. Pop
National Institute for Research and Development in Informatics (ICI), Bucharest, Romania

B. Di Martino et al. (eds.), *Internet of Everything*, Internet of Things,
https://doi.org/10.1007/978-981-10-5861-5_10

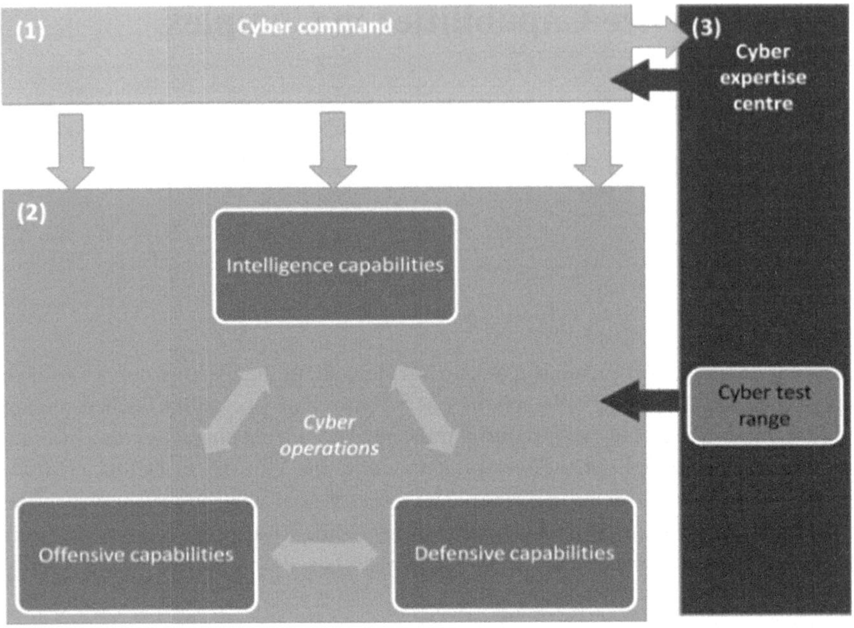

Fig. 1 MoD future governance framework structure

The Problem Area
In its plans for cyber training and defence, the Ministries of Defence of several countries considered its massive cost reduction for its operation, and its desire was getting towards the field of digital resilience and cyber operations [1]. There are some examples of governments, like UK Government and Netherlands Government, that dedicated around €50 million to invest in the field of digital resilience and cyber operations to be used to reinforce the kinetic weapon arsenal in 2016.
From the US Government perspective, that can be generalized, the strategy of a well-prepared MoD's cyber component has six objectives [2], presented in Fig. 1:

1. realize a cohesive approach and some good assessments from a cyber security point of view;
2. increase the cyber security resilience of the MoD and other critical infrastructures;
3. development the MoD's capabilities to execute cyber operations (both offensive and defensive);
4. develop more intelligence capabilities in the cyber domain;
5. develop knowledge and acquire new and innovative capabilities in the cyber security field;
6. develop national and international cooperation with another MoD's or CSIRTs.

The chapter is organized as follow. Section 2 presents the data manipulation challenges being focused on spatial and temporal databases, key-value stores and no-SQL, data handling and data cleaning, Big Data processing stack and processing techniques. Section 3 describes the reduction techniques: descriptive analytics, predictive analytics and prescriptive analytics. In Sect. 4 we present a case study focused on CyberWater, which is a research project aiming to create a prototype platform using advanced computational and communications. The future governance framework structure of MoD will be the one above [3]. The first entity in the structure is Cyber Command which will take over the cyber operations. The second entity will be cyber operations that contains the intelligence capabilities, defensive capabilities and offensive capabilities. The last entity is the Cyber Expertise Center that concentrates on the skills and the knowledge regarding the cyber operations in the MoD. This entity then will provide a cyber test range (CTR).

The cyber test range will be one of the functionalities that supports cyber operations. A CTR can be considered as a "cyber shooting range", comparable to a shooting range in the physical world, where military personnel can perform offensive and defensive training and test their other skills.

The cyber security perspective from a MoD with the CTR, can be easily applied in the law enforcement cyber departments, but its function is not obviously determined in terms of goals, objectives or specifications.

The current IT test environments in the MoD are not suitable for the CTR used in the cyber operations because these environments are used for the availability and capacity testing purposes as part of IT Infrastructure Library or Service Management processes [4].

The Research Goal

The research main goal is to design a roadmap for the development of a cyber test range.

Following are some sub-goals that are derived from the main goal:

1. Delivering the definition for cyber operations, its capabilities and to establish the activities that are conducted within cyber operations capabilities.
2. Acquiring knowledge about the use and developments in cyber test ranges and to provide CTR business functions.
3. Determining the CTR business functions that support offensive, defensive and intelligence capabilities. Business functions are a series of logically related services performed together to obtain a defined set of results.
4. Identifying the technical and organizational requirement s for delivering the CTR business functions.
5. Designing a timeline for the implementation of business functions and the technical and organizational requirements needed to deliver business functions. The roadmap delivers the necessary input for the change management for the implementation of a CTR.

Research Scope

NATO uses the scale of DOTMPLFI (Doctrine, organization, training, material, leadership and education, personnel, facilities and interoperability) to determine the functionality of any capability. It is said before the current IT environments in MoD are not apt for the CTR, so that scale should be redefined, redesigned, developed, and re-implemented.

This study concentrates on the organizational (including personnel) and technical requirements. Without good organization to keep the CTR there is no CTR function. The other requirements in the scale will be supportive for the organizational and technical.

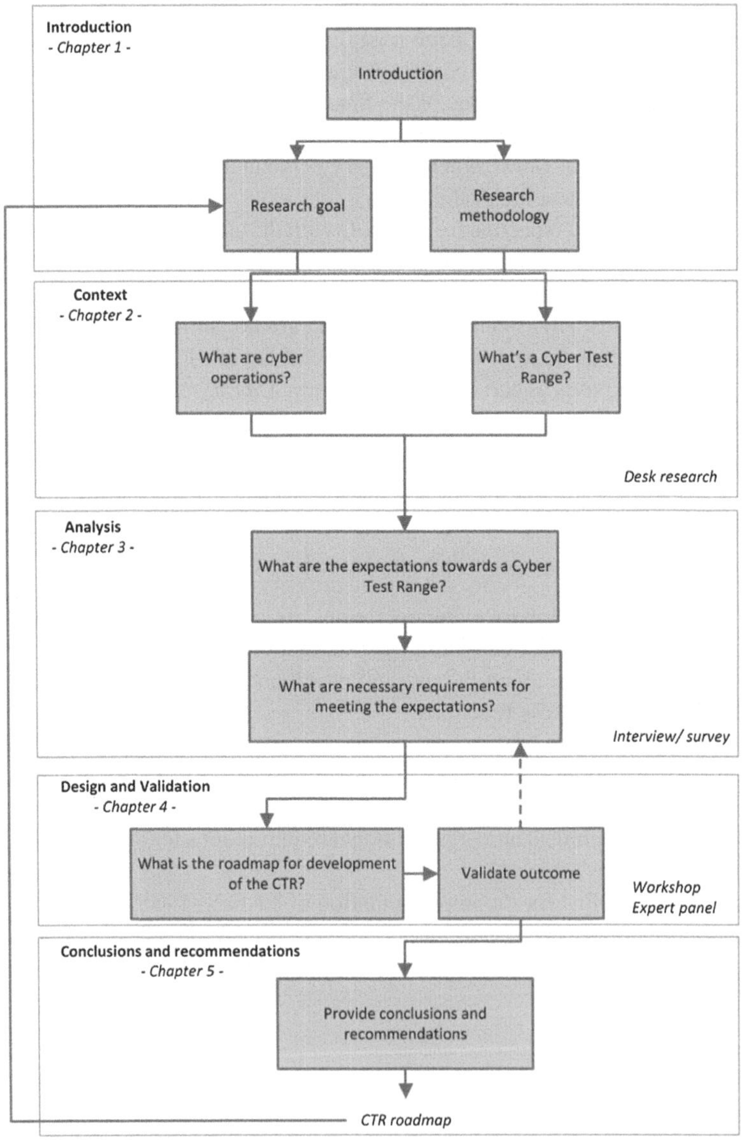

Fig. 2 Research methodology

The Research Methodology

The research is conducted through the following methodology (that can be grouped in five chapters of work, presented in Fig. 2):

1. Understand the main idea and the need of cyber operations and cyber test ranges.
2. Perform an in-depth analysis to determine the functions and requirements applicable to the MoD CTR.
3. Design a complex roadmap for the cyber test range based on the result of the analysis phase.

2 Cyber Operations

Cyber (space) operations are defined as "the employment of cyberspace capabilities where the main purpose is to achieve military objectives or effects in cyberspace or through it".

NATO uses the following definitions to describe the capabilities within Cyber Operations (see Fig. 3):

- Computer Network Operations (CNO)—Computer Network Operations (with three components Computer Network Attack, Exploitation, and Protection) focused to obtain unrestricted access to computer networks to disrupt or deny their capabilities, or use them like a bot.
- Computer network Defence (CND)—Actions to protect against denial or destruction of information located in computers, computer networks or the networks themselves.
- Computer network attack (CNA)—Action taken to deny/destroy information from computers, computer networks.
- Computer network exploitation (CNE)—Action taken to make use of a computer or computer network and the information located on them, to gain advantage.

Cyber operations are conducted through intelligence, offensive and defensive capabilities. The Cyber Defence aims at protecting own networks and systems. The Cyber offence aims at disrupting, denying, degrading, or destroying networks and systems. The Cyber intelligence enables intelligence collection through networks and systems [5, 6].

The activities conducted in the cyber-attacks (offensive) and intelligence are similar and they aim at accessing the system to lead to a planned effect. These activities consist of: recon, scan, access, escalate, exfiltration, assault, sustain, and obfuscate.

The activities conducted in the cyber Defence follow the life cycle of an incident and consist of six main activities that are part of the NATO Framework: malicious activity detection, attack termination, prevention or mitigation, dynamic risk damage or attack assessment, cyber-attack recovery, timely decision making and the cyber defence information management.

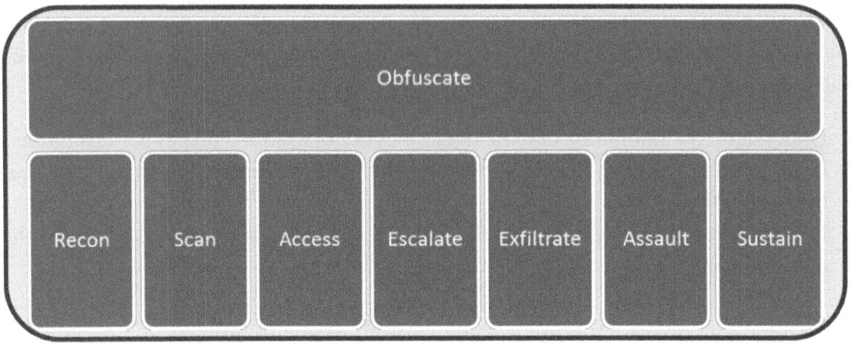

Fig. 3 Cyber operations capabilities

Those activities are well explained in the figure below, and they are part of the Cyber Defence Capability Framework released by NATO in 2010 (see Fig. 4), which had the main goal to provide NATO and its Nations a well-documented standard for cyber operations, to develop a better multinational cooperation in the cyber security field, to coordinate cyber defence activities.

Developments in Cyber Test Ranges

Cyber test Ranges (CTR) are defined as not-real (virtual) environments used for research, development, evaluation, and training purposes within the domain of the cyber. The aim of the test ranges involves recreating real world situations but without any harm to the real-world networks. According to the military views, the CTR can be used to defend and attack infrastructures or military capabilities.

CTR requirements are demanding. They should replicate the networks and computer systems, imitate the business operations, and produce generate realistic traffic to conduct tests without harming the real environments. And they should be flexible to adapt their configuration with other test ranges for supporting the large-scale experiments or exercises.

From military point of view, CTR can be understood as an environment that offers partners the capacity to—even more successfully - guard and assault (or gain intelligence information about) cyber critical infrastructures or military capacities.

Cyber comprises of numerous components that also are being seen in an unexpected way [7]. Cyber component is putted in a complex network that is well resumed in the following picture.

Case Studies

There is a various number of CTR that have been made operational or are still under implementation. These CTR's are good to extract the current or future characteristics and objectives, and in this manner, to add to a superior comprehension of how are cyber test ranges are designed for training purposes to fight against cybercrimes and cyberterrorism.

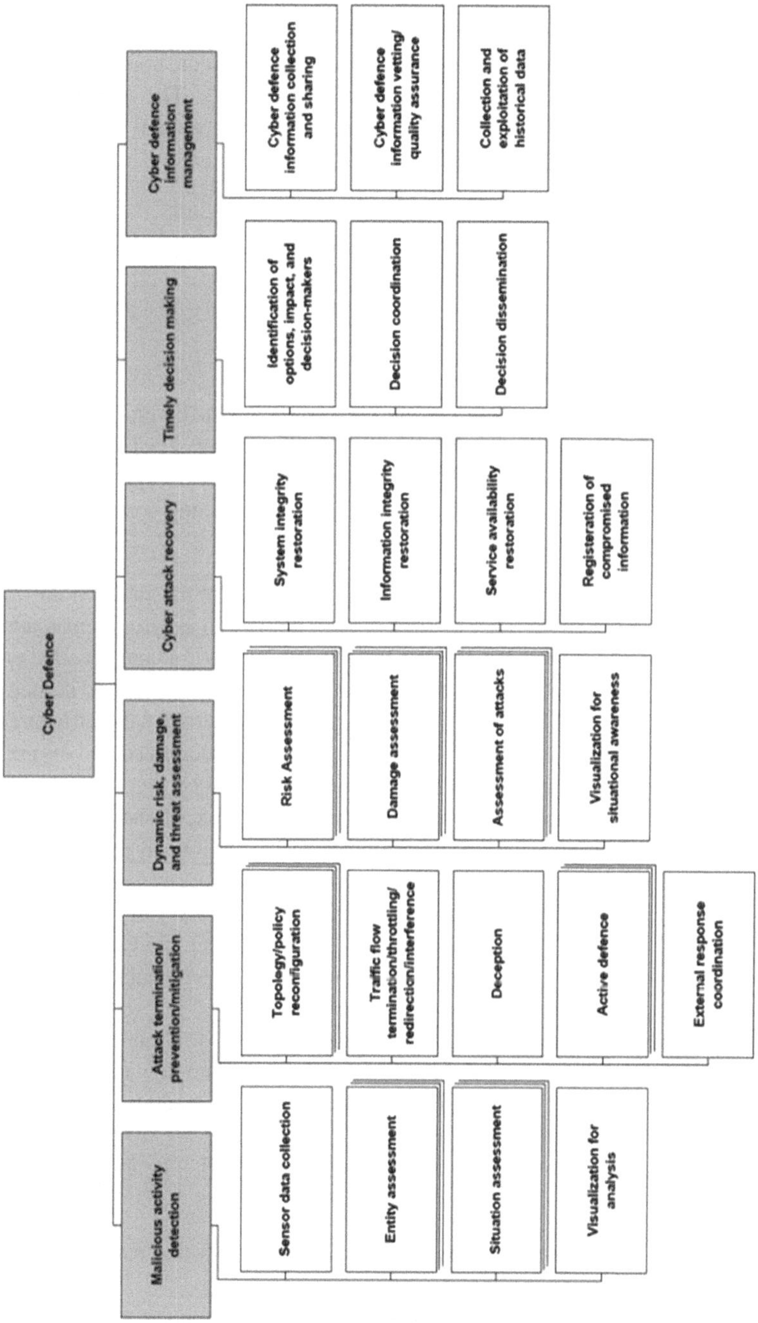

Fig. 4 Cyber defence capability framework

Examples:
1. The United States CTR—The US is in the phase of implementing a National Cyber Test Range (NCR). This cyber range [8] will provide the infrastructure and software tools for a secure testing capability to rapidly emulate large-scale complex networks that simulate the depth and diversity of real-world networks. The implementation started in 2008 and will service [9] both researchers and operational users.

Experimental Researchers will have:

- the capacity to quantify the progress of their analysis in detail;
- the appropriate classified or unclassified environment;
- experiments against sensible threats;
- the use of investigative approach to track and trace examinations and results.

Operational users will have:

- proper test and assessment of military and government net-driven frameworks or systems to guarantee current and future guard against cyber attacks;
- fast evaluation of the Nation's current and future cyber research programs;
- cyber security experimentation technologies for all ranges and communities;
- decreased time/cost for cyber tests.

In addition to NCR, the US cyber experts started developing in 2006 an information operations (IO) range [10]. Its goal was to deliver an environment made of procedures and structures which set up a sensible test, preparing, and practice environment for creating and operationalizing IO capabilities and their related strategies, methods, and procedures. The IO range go in this way speaks to actual battle targets, frameworks, and circumstances, permitting clients to lead specialized and execution affirmation testing for IO capacity framework certification [11].

2. NATO—NATO Cooperative Cyber Defence Centre of Excellence (NATO CCD COE) runs a cyber lab, as stated by the Director of CCD COE in an email correspondence. The cyber lab aimed at operational users in support of technical courses for training [12] and technical expertise. In this range are developed two important exercises—Lock Shield (red team vs blue team exercise) and Cyber Coalition [13] (exercise based on different scenarios that involves malware analysis, host and network forensics, traffic analysis and reporting) [14].

3. The United Kingdom opened its cyber range in 2010 [15]. Their CTR "is able to simulate a large infrastructures and global threats and evaluate how these networks, whether military, civilian or commercial, respond to an attack to develop capabilities that will make these networks more secure".

Northrop Grumman delivers the test range facilities [16]. The cyber range has four common uses:

- Training aimed at preventing falling victim to cyber-attacks and response training aimed at improving the handling of cyber-attacks.
- To getting and understanding of the robustness of the IT-architecture and to understand the consequences of additions or changes to the IT-architecture

- To test and to benchmark IT-components.
- Research and development.

This Federated Cyber Range (FCR) [17], as it is called, is intended to permit interoperability with other digital reaches to empower huge scale explores past the extent of a solitary office beyond the scope of a single facility [18].

3 MoD Cyber Test Range

The expectations of the MoD business towards the CTR include the CTR business functions that consist of many levels. The first level relates between the CTR business functions with the cyber operations: the functions ease the execution of the cyber operations. The second level is about specific business functions that support one of the capabilities in the domain of cyber operations: defence, offense, or intelligence.

The generic business functions are business functions that support the daily operations and the enable the research and the development. The CTR, to support the operations, try to present business functions that help the personnel to act and assess effectiveness of the current capabilities in the cyber domain. Also, the CTR, to enable the research and the development, attempt to present business functions that help the researchers to carry out the researches into future cyber solutions and to research more when the external solutions enrich the MoD.

The specific functions [19] are meant to support one of the three capabilities. To illustrate that, one of the generic business functions is to enable the personnel to act in the cyber domain; this function can be specified into training the personnel to endure the cyber-attacks in the case of cyber defence. Moreover, these specific functions can be specified into services and the CTR service components. This advantage of specification makes the activities supported by the CTR in one of the three capabilities. That means the CTR grant an added value for each capability and each activity.

The technical and organizational requirements are crucial for the delivery if the CTR business functions (see Fig. 5). The technical requirements include the ability in designing IT environments but they should be scalable in assets and flexible in positioning the configurations. Also, the security requirements are asserted. The organization requirements include IT-staff that preserve and organize the CTR and staff that attend the training and the experiments [20].

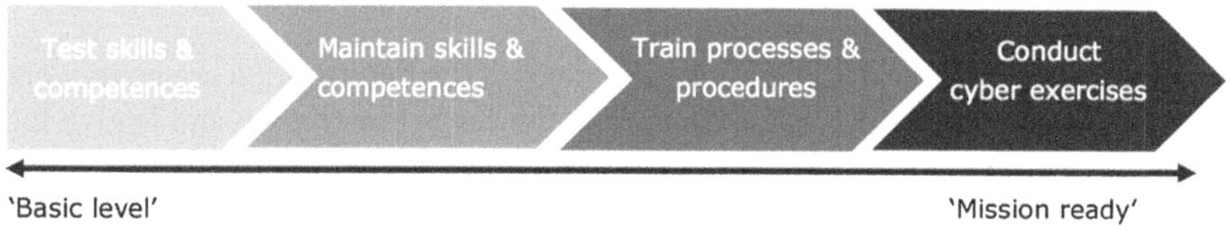

Fig. 5 CTR business functions

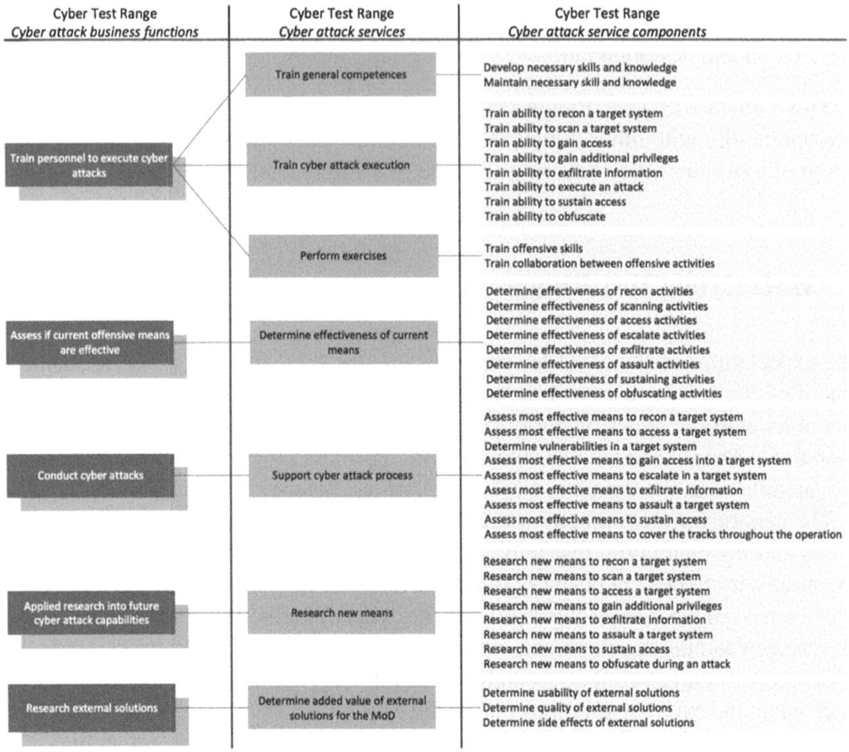

Fig. 6 Cyber test range—Offensive business functions

Exercises are a critical component in cyber security operations as it consolidates each part of cyber operations into a close genuine live action. The three capabilities can also train each other when taking part in an integral exercise with red teams (attackers) and blue teams (defenders), such as LockShield organized every year in Estonia by CCDCOE (NATO).

Cyber defence/attack—describes the desires towards the CTR from a defensive/offensive point of view. The detailed overview of the cyber Defence expectations consists of three components [21]:

1. The business capacities went for supporting cyber defence/attacks.
2. A further specification of the capacities into CTR administrations pointed at supporting cyber defence/attacks.
3. A breakdown of the CTR administrations into CTR administration segments went for supporting cyber defence/attacks. The figures below (Figs. 6 and 7) give the realistic breakdown of the CTR desires from a Defence and offensive point of view.

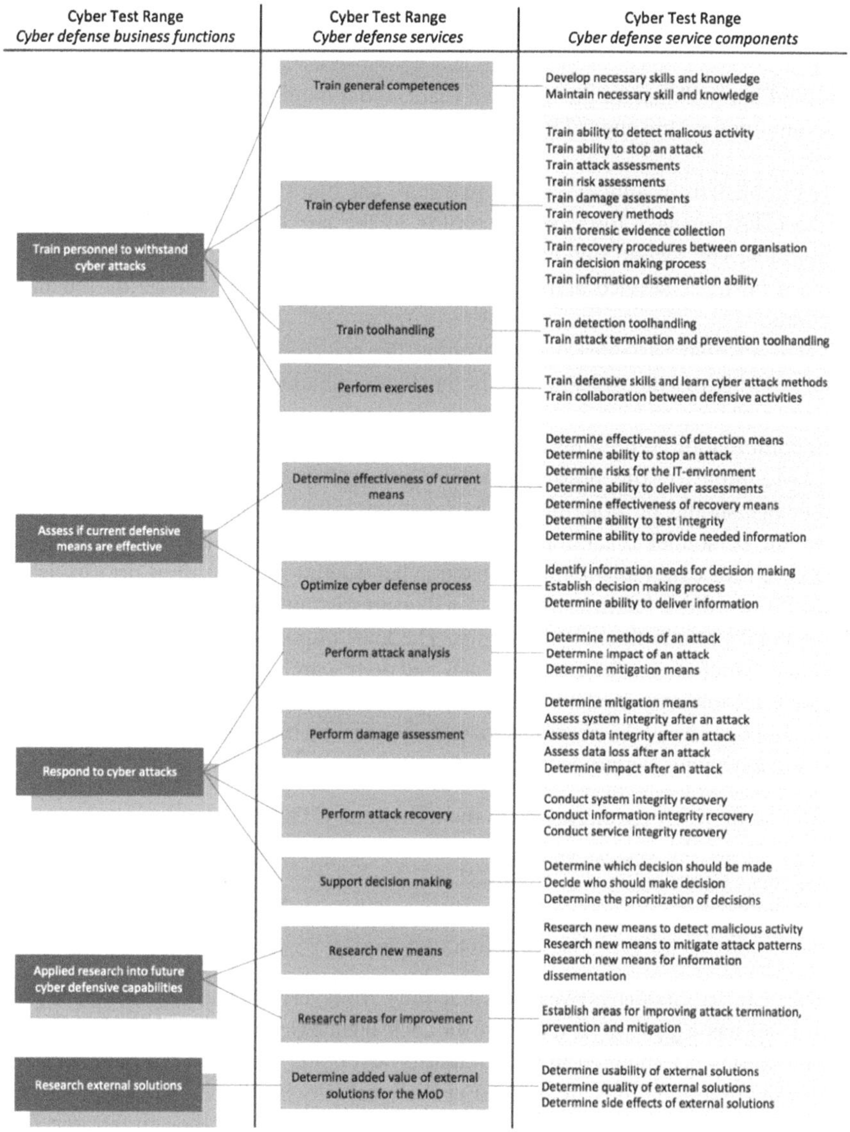

Fig. 7 Cyber test range—defensive business functions

4 Roadmap for the Cyber Test Range

This roadmap will last for the next five years, and it includes the delivery of the business functions within the CTR and the needed technical and organization requirements.

To establish the roadmap requires two step. The first step is to identify the priority for the CTR business functions according to the perspective of cyber operations capabilities. The second step is to determine the various levels of the functionalities within a business service and the necessary requirements to deliver the business functions.

There are two variables that determine the priorities. The first variable is the urgency (which is the need to use the bushiness quickly) and the second variable is the complexity (which is to know the necessary requirements). Both variables (their combination) represents the priorities for the business functions; high urgency and low complexity functions should be conducted first, but low urgency and high complexity should be done next. The most important priority is the function the enable the personnel to act in the cyber domain [22]. The next priority is the ability to research for external solutions for the cyber operations. The third priority is the ability to use the CTR in response to cyber-attacks and use the CTR to carry out cyber-attacks or intelligence operations. The least important priority is the business functions which assess the current means and do research into future cyber solutions.

The CTR maturity model is designed to define the different eves of the functionalities and determine the necessary requirements. The methodology for defining the model test depends on three steps [23]:

1. To link the CTR requirements to the individual CTR services.
2. To abstract the requirements related to the CTR services to the level of CTR business function.
3. To divide the requirements into different levels and link these requirements to service levels for each business function.

The maturity model has five levels that range from very basic to very advanced [24]. Each level has a general description. In each business function, the description of functionalities is explained in each maturity level, and the needed requirements to deliver the functionalities are defined in each maturity level.

In general, the roadmap presents the business functions and their timeframe. The business functions start with ambition level and end with maturity level.

Also, it offers the requirements that deliver the business functions based on their ambition levels [25].

More details are presented in Fig. 8.

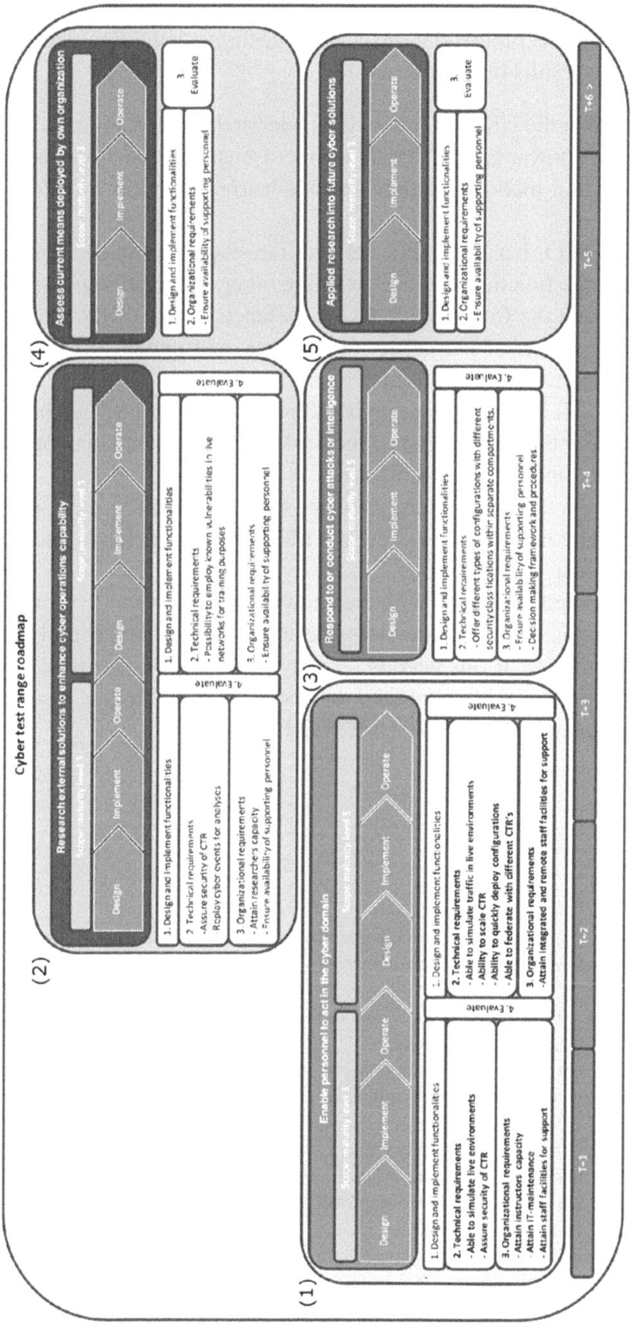

Fig. 8 The cyber test range roadmap

5 Conclusions and Recommendations

Following are some recommendations and suggestions that the researcher came up with based on the results of the study.

1. Collaborate with the UK MoD about the Federated Cyber Range. Validate the NL MoD approach towards the CTR in terms of business functions, requirements, and roadmap and incorporate their lessons learned into the NL MoD approach for the CTR.
2. Work with NATO, because there are also developments in the realization of a 'cyber test range functionality and examine other cooperation possibilities.
3. Collaborate with the Cooperative Cyber Defence Centre of Excellence because they are very expert at preparing, facilitating, and conducting cyber Defence exercises supported by a cyber lab.
4. Develop the CTR business function based on the descriptions and the CTR maturity levels under the supervision of the Taskforce Cyber and in cooperation with the three cyber operations capabilities.
5. Acquire through the forthcoming Defence Cyber Expertise Centre the resources (researchers and instructors) to conduct trainings and exercises and do research and development.
6. Define the research questions with the knowledge institutions to research the possibilities for replicating live networks, the ability for organizing real live environments used in the CTR, the needed security to protect the sensitive information, the risk and heath management, and the access from difference locations.
7. Define, design, and develop the DOTMPLFI measures.

References

1. Ottis, Rain and Lorents, Peeter. 2010. Cyberspace: Definitions and Implications. In *5th international conference on information warfare and security*, (Dayton OH, US: Cooperative Cyber Defence Centre of Excellence).
2. Wiener, Norbert. 1948. *Cybernetics: Or control and communication in the animal and the machine*. Cambridge: The MIT Press.
3. Thill, Scott. March 17, 1948: William Gibson. 2011. Father of Cyberspace. Wired.com. March 2011. http://www.wired.com/thisdayintech/2011/03/0317cyberspace-author-william-gibson-born/.
4. Kuehl, Dr Dan. 2009. From Cyberspace to Cyberpower: Defining the Problem. [book auth.] Stuart H. Starr, and Larry K. Wentz Franklin D. Kramer. Cyberpower and National Security. s.l.: Potomac Books, Inc., Vols. in Cyberpower and National Security, ed. Franklin D. Kramer, Stuart H. Starr, and Larry K. Wentz.
5. Cornish, Paul, David Livingstone, Dave Clemente, and Claire Yorke. 2010. *On cyber warfare*. London: Chatham House. November.
6. Andress and Winterfeld. 2011. *Cyber warfare; techniques, tactics and tools for security practitioners*. New York: Elsevier.

7. US Department of Defence. The National Military Strategy for Cyberspace Operations. December 2006.
8. Ministry of Security and Justice. Cyber Security Beeld Nederland. December June 2012. CSBN-2.
9. US Department of Defence. Joint Publication 3-0, Joint Operations. August 2011.
10. NATO. Allied Joint Doctrine for Information Operations. November 2009. AJP 3.10.
11. Mirkovic, Jelena, et al. 2010. The DETER project: Advancing the science of cyber security experimentation and test. IEEE. 978-1-4244-6048-9/10.
12. West-Brown, et al. *Handbook for computer security incident response teams (CSIRTs)*. (New York: Carnegie Mellon University, 2003). CMU/SEI-2003-HB-002.
13. Benzel, et al. 2007. Design, Deployement and Use of the DETER Testbed. In *DETER community workshop on cyber-security and test*, Boston, August 2007.
14. NC3A. 2010. Cyber Defence Capability Framework. December 2010.
15. BuxBaum, Peter A. 2011. Building a Better 'Cyber Range'. August 2011.
16. Sabo, Robert P. 2006. Standing Up the Information Operations Range. 2006.
17. Powell, Robert, Holmes, Timoty K. and Pie, Cesar E. 2010. The information assurance range. *ITEA Journal* 31: 473–477.
18. UK Ministry of Defence. Defence Minister opens UK cyber security test range. Ministry of Defence. http://www.mod.uk/DefenceInternet/DefenceNews/DefencePolicyAndBusiness/DefenceMinisterOpensUkCyberSecurityTestRange.htm.
19. US Department of Defence. 2009. The Global Information Grid (GIG) 2.0. Concept of Operations. March 2009. Version 1.1.
20. Watson. Combat Readiness through Resilience in Hostile Cyber Environments.
21. Welshans. 2010. History of cyber testing and evaluation - A voice from the front lines. *ITEA Journal* 31: 449–452.
22. Benzel, et al. 2009. Current Developements in DETER Cybersecurity Testbed Technology.
23. DARPA. National Cyber Range. http://www.darpa.mil/Our_Work/STO/Programs/National_Cyber_Range_(NCR).aspx.
24. Defence Information Systems Agency. Department of Defence Information Assurance Range: A Venue for Test and Evaluation in Cyberspace. August 2011.
25. NATO Cooperative Cyber Defence Centre of Excellence. CCD COE Training Courses -CCD COE. http://www.ccdcoe.org/236.html.

The manufacturer's authorised representative in the EU is Springer Nature Customer Service Centre GmbH, Europaplatz 3, 69115 Heidelberg, Germany. If you have any concerns regarding our products, please contact ProductSafety@springernature.com

Printed and bound by CPI Group (UK) Ltd, Croydon, CR0 4YY
30/09/2026
02225440-0003